134
Advances in Biochemical Engineering/Biotechnology

For further volumes:
http://www.springer.com/series/10

Aims and Scope

This book series reviews current trends in modern biotechnology and biochemical engineering. Its aim is to cover all aspects of these interdisciplinary disciplines, where knowledge, methods and expertise are required from chemistry, biochemistry, microbiology, molecular biology, chemical engineering and computer science.

Volumes are organized topically and provide a comprehensive discussion of developments in the field over the past 3–5 years. The series also discusses new discoveries and applications. Special volumes are dedicated to selected topics which focus on new biotechnological products and new processes for their synthesis and purification.

In general, volumes are edited by well-known guest editors. The series editor and publisher will, however, always be pleased to receive suggestions and supplementary information. Manuscripts are accepted in English.

In references, Advances in Biochemical Engineering/Biotechnology is abbreviated as *Adv. Biochem. Engin./Biotechnol.* and cited as a journal.

Pauline M. Doran
Editor

Biotechnology of Hairy Root Systems

With contributions by

Z. Al-Shalabi · A. Araujo · V. P. Bulgakov
J. A. Chichester · P. M. Doran · V. Gopalasamy-Raju
T. Y. Gorpenchenko · C.-Z. Liu · V. Mett
K. Musiychuk · G. A. Palmer · P. de la Rosa · K. Ruby
G. W. Sander · J. V. Shanks · Y. N. Shkryl
M. Skarjinskaia · A. R. Stiles · S. J. Streatfield
K. Taylor · N. Ugulava · G. N. Veremeichik
Y. V. Vereshchagina · J. W. Wang · J. Y. Wu
V. Yusibov · L. Zhao

Springer

Editor
Pauline M. Doran
Faculty of Life and Social Sciences
Swinburne University of Technology
Hawthorn, Melbourne, VIC
Australia

ISSN 0724-6145 ISSN 1616-8542 (electronic)
ISBN 978-3-642-39018-0 ISBN 978-3-642-39019-7 (eBook)
DOI 10.1007/978-3-642-39019-7
Springer Heidelberg New York Dordrecht London

Library of Congress Control Number: 2013943946

Printed on acid-free paper

Springer is part of Springer Science+Business Media (www.springer.com)

Preface

Hairy roots were studied in the 1930s–1960s as an indicator of pathogen attack in horticultural plants such as apple trees and roses. The responsible bacterial agent, *Agrobacterium rhizogenes*, was identified and the role of gene transfer from the bacterial Ri (root-inducing) plasmid was revealed in the 1970s and 1980s. Starting in the mid-1980s, it was recognised that hairy roots cultured in vitro could be used as a tool in studies of secondary metabolism and, ultimately, as production systems for natural products. This led to a burst of research activity in these areas, focusing on the intrinsic benefits of root organ differentiation for secondary pathway activity. This research is ongoing today. In the meantime, a greater and more detailed understanding of the molecular biology of bacteria–plant infection emerged, supporting an expansion in the range of applications of hairy roots in biology, biotechnology and engineering. Hairy roots became a convenient and reliable tool for many laboratory-based studies.

In 1997, a research monograph, *Hairy Roots: Culture and Applications* (P. M. Doran, Harwood Academic) provided a snapshot of the state of research activity in hairy root biotechnology at that time. In 21 chapters, the laboratory protocols for initiation, culture and genetic transformation of hairy roots were set out, applications in alkaloid synthesis and plant propagation were described, and bioprocessing considerations for large-scale hairy root culture were outlined. Early work on the expression of foreign proteins and metal uptake using hairy roots was also represented. This volume reflected a broad cross-disciplinary interest in hairy roots, involving scientists in many areas ranging from genetics and molecular biology to horticulture, medicine, environmental studies and bioprocess engineering.

Today, the exploitation of hairy roots in biotechnology and bioprocessing research continues, with over 100 papers published each year on the topic. This special issue provides an update in those areas where our knowledge on hairy roots and the development of their applications have advanced significantly in recent times. The complex effects that agrobacterial genes have on plant cells, such as perturbation of stress responses and defence strategies, are now recognised as important mechanisms of pathogenicity in infected plants. Increasing interest in secondary pathway elucidation and the use of metabolic engineering principles to improve pathway function has led to an expanded role for hairy roots in this context. Further progress to elicit and enhance secondary metabolite production in bioprocessing systems has been made. New products synthesised using hairy roots, such as metallo-organic nanocrystals and foreign proteins for vaccine synthesis, are also highlighted in this volume as an encouragement for further work in these areas.

I would like to thank all the authors who have contributed chapters to this special issue for their readiness to be involved, their expertise, and their efforts in creating comprehensive review papers. I am also grateful to Thomas Scheper and Springer for their friendly collegiality and support of this project.

Pauline M. Doran

Contents

Contents

Adv Biochem Eng Biotechnol (2013) 134: 1–22
DOI: 10.1007/10_2013_179

Published Online: 11 April 2013

Recent Advances in the Understanding of *Agrobacterium rhizogenes*-Derived Genes and Their Effects on Stress Resistance and Plant Metabolism

Victor P. Bulgakov, Yuri N. Shkryl, Galina N. Veremeichik, Tatiana Y. Gorpenchenko and Yuliya V. Vereshchagina

Abstract It is commonly accepted that the plant pathogens *Agrobacterium rhizogenes* and *Agrobacterium tumefaciens*, acting via their T-DNA oncogenes, disturb hormone metabolism or hormone perception pathways in plants, thereby attaining their aim of successful pathogenesis. In this work, we summarize recent data on the *A. rhizogenes rolC* and *rolB* oncogenes in comparison to the *A. tumefaciens 6b* oncogene with respect to their effects on the physiology of transformed cells. The newly discovered functions of the *rol* genes include the modulation of secondary metabolism, the modulation of levels of intracellular ROS and stress resistance of transformed cells, changed sucrose metabolism, and the inhibition of programmed cell death. We show that the *rol* genes do not have suppressive effects on plant innate immunity; rather, these genes activate plant defense reactions. The existence of not only the hormone-related mechanism of pathogenicity but also the defense-related mechanism of pathogenicity during plant-*Agrobacterium* interactions is suggested.

Keywords *6b* gene · *Agrobacterium* · Plant defense reactions · Plant oncogenes · *rol* genes · Secondary metabolism

Abbreviations

ROS	Reactive oxygen species
siRNA	Small interfering RNA

V. P. Bulgakov (✉) · Y. N. Shkryl · G. N. Veremeichik ·
T. Y. Gorpenchenko · Y. V. Vereshchagina
Institute of Biology and Soil Science, Far East Branch of the Russian Academy of Sciences, Vladivostok 690022, Russia
e-mail: bulgakov@ibss.dvo.ru

V. P. Bulgakov
Far Eastern Federal University, Vladivostok 690950, Russia

miRNA	MicroRNA
RISC	RNA-induced silencing complex
dsDNA	Double-stranded DNA
H_2DCF-DA	2,7-dichlorofluorescein diacetate

Contents

1 Introduction

Agrobacterium tumefaciens and *Agrobacterium rhizogenes* plant pathogens contain Ti and Ri plasmids, which induce crown galls and hairy roots, respectively, during plant-*Agrobacterium* interactions. T-DNA genes of these plasmids have been found to influence tumor or hairy root development. Many T-DNA genes belong to a highly diverged family defined on the basis of weak and partial protein homologies. This so-called *plast* [40] or *rolB* [48] family includes 17 subfamilies. Although *A. tumefaciens* and *A. rhizogenes* are related micro-organisms, the structural and functional divergence of their T-DNA oncogenes is obvious. The uniqueness of these plant pathogens is that they use *plast* genes to adapt the environment for their benefit.

T-DNA of Ri plasmids includes the genes encoding oncoproteins RolA, RolB, RolC, RolD, open reading frame (Orf) 8, Orf13, and Orf14. T-DNA of Ti plasmids contains genes encoding oncoproteins Lso, 3′, 5, 6a, 6b, 7, B, C′, and others. The most-studied proteins of the RolB family are RolB, RolC, and 6b. According to the phylogenetic tree of 50 RolB-like proteins [34], RolC is more closely related to Orf14 and Orf13 than to RolB. RolC is more closely related to 6b than to RolB.

Because the *rol* genes from Ri plasmids play a key role in the formation of hairy roots (genetically transformed roots), they have been extensively studied during the past 30 years. Presently, hairy roots are considered to be important producers of bioactive molecules, such as secondary metabolites and recombinant proteins [28, 32, 63]. Genetically transformed roots represent a valuable model for phytoremediation and virus propagation studies [20, 47, 53]. The *rol* genes are used for engineering high yields of secondary metabolites [10, 12] and for the generation of compact growth in ornamental plants [16, 17, 41]. In this work, we discuss recent findings regarding RolB, RolC, and 6b proteins, which possess very complex biological activity and have great potential in plant metabolic engineering and biotechnology.

2 New Functions of the *rolB* and *rolC* Genes in Plant Cells

2.1 ROS Homeostasis and Expression of Antioxidant Genes

The first indication that the expression of individual *rol* genes in plant cells has an effect on the metabolism of reactive oxygen species (ROS) was published in 2008 [11]. This investigation was performed to determine whether the increased production of secondary metabolites in *Rubia cordifolia* cultured cells is associated with the oxidative burst. It was known that fungal elicitors and other microbe-associated molecular patterns trigger a rapid ROS accumulation (oxidative burst), with subsequent activation of various defense systems [59]. As *rolC* expression led to activation of secondary metabolites [4, 5, 6, 49, 54] and pathogenesis-related proteins [37], the activation of ROS production by the oncogene could be expected. In contrast, earlier pharmacological studies suggested that ROS are not involved in the activator function of *rolC* [7].

ROS measurements in *rolC*-expressing *R. cordifolia* callus cells were performed by laser confocal microscopy at a single-cell level (calli were cultivated in a liquid medium to obtain single cells and small cell aggregates). The results indicated decreased intracellular ROS levels in *rolC*-expressing cells compared with control cells [11]. Moreover, the gene attenuated ROS accumulation after ROS-inducing treatments. A similar behavior has been demonstrated for the *rolB* gene [14]. The effects of single *rol* genes and their combinations on ROS levels are presented in Table 1. In this table, we have included both published results [11, 14, 55] and unpublished results to provide a complete picture of effects.

ROS inhibition was most strongly induced by *rolC,* followed by *rolB* and *rolA*. The combined action of all three genes (*rolABC* construct) caused a moderate decrease in ROS. Obviously, the *rol* genes do not act synergistically on the decrease in ROS. The behavior of calli transformed with the wildtype A4 plasmid was similar to that of calli transformed with the *rolABC* construct (Table 1). The *rolABC* calli and pRiA4 calli effectively sustained acute elevations of ROS [55].

Table 1 Individual and combined expression of the *rol* genes in suspension-cultivated *R. cordifolia* cells decreases ROS levels

Intracellular ROS level	Nontransformed control	*rolA*	*rolB*	*rolC*	*rolABC*	*pRiA4*
Steady-state conditions	100 ± 3	93 ± 3[a]	78 ± 4 [14]	56 ± 6 [11]	90 ± 3 [55]	80 ± 4 [55]
ROS-inducing treatment	160 ± 4 [11, 14]	109 ± 5[a]	96 ± 5 [14]	74 ± 5 [11]	101 ± 4[a]	113 ± 4[a]

The data are presented as percentage concentrations of intracellular ROS. The basic level of ROS in control nonstressed cells has been accepted for 100 %. The 5-day-old cell suspension cultures were loaded with H_2DCF-DA and analyzed by confocal microscopy as described previously [11]. In the ROS-inducing treatment, *R. cordifolia* cells were grown for 4 days in the dark and treated with paraquat (Aldrich, 10 μM, final conc.) for 1 h under continuous light exposure (200 μmol m^{-2} s^{-1} radiation). The data were obtained in three independent experiments and presented as mean ± SE

[a] Our unpublished results

For example, upon treatment with the ROS-inducing herbicide paraquat (in the presence of light, paraquat reduces oxygen to a superoxide radical, which subsequently dismutates to H_2O_2), these calli maintain normal ROS levels, whereas the control calli increase intracellular ROS levels up to 170 % (Table 1). In agreement with our data, decreased ROS levels were recently found in hairy roots of *Nicotiana tabacum* [45]. In this work, the authors detected a 50 % decrease in ROS in several tobacco hairy root cultures compared with nontransformed roots. Both strains of *A. rhizogenes* studied, 15834 and A4, induced the effect.

Notably, *rolB* has a definitive differential effect on ROS metabolism compared with *rolC*. In long-term experiments, *rolC* did not prevent the toxic effect of paraquat on cells, whereas *rolB* prevented cell death [14]. Our data indicate that some components of the cellular ROS-detoxifying system tend to be suppressed by *rolC*. Expression of genes encoding ascorbate peroxidases, catalases, and superoxide dismutases was either not changed or decreased in *rolC* cells compared to normal cells (Fig. 1b–d). In contrast, *rolB* activates all these defensive systems. The protective effect of *rolB* on excessive ROS accumulation was so powerful that *rolB*-transformed cells of *R. cordifolia* could tolerate superlethal doses of the ROS-inducing inhibitor menadione [14]. This protective effect is a long-lasting one. From the physiological point of view, the effect of *rolB* is similar to the phenomenon known as stress acclimation or, more specifically, systemic acquired acclimation [27, 44]. During ROS-induced stress acclimation, plants produce catalases, ascorbate peroxidases, and other ROS-detoxifying enzymes to protect their cells against new stresses [27]. This leads to sustained antioxidant defenses and the protection of the plants from subsequent stresses.

The *rolC* gene works better during short-term stress treatments and does not protect cells against permanent stress. The combined action of the *rol* genes with other T-DNA oncogenes resulted in a moderate activation of the ROS-detoxifying genes in pRiA4-transformed calli [55] and of ROS-detoxifying enzymes in hairy roots [45].

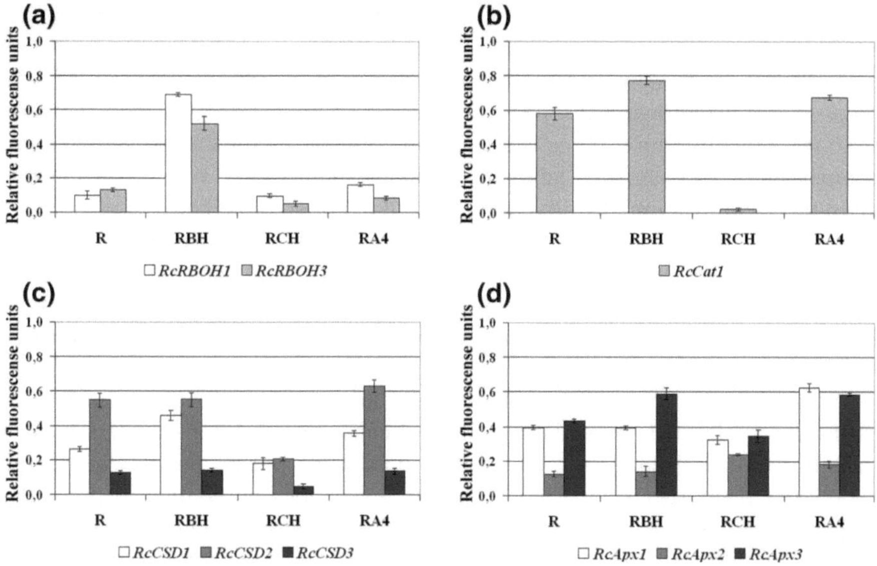

Fig. 1 Expression of genes participating in ROS generation (**a**) and ROS detoxification (**b**–**d**) in *R. cordifolia* callus cells. *R* nontransformed control, *RBH* high-*rolB*-expressing cells, *RCH* high-*rolC*-expressing cells, *RA4 A. rhizogenes* A4 transformed cells. *RcRBOH1* and *RcRBOH1* are the NADPH oxidase genes (respiratory burst oxidase homologues), *RcCat1* is a catalase gene; *RcCSD1*, *RcCSD2*, and *RcCSD3* are the Cu/Zn superoxide dismutase genes; *RcApx1*, *RcApx2*, and *RcApx3* are the ascorbate peroxidase genes. RNA samples were isolated from callus cultures during the linear phase of growth (20–22 days). Quantitative real-time RT-PCR (qRT-PCR) was performed as described previously [55]. The qRT-PCR analysis was performed using the Bio-Rad CFX96 Real-Time System (Bio-Rad Laboratories, Inc., USA) with a 2.5 × SYBR green PCR master mix. The *RcActin* gene (GenBank accession no. DQ531565) was used as a reference gene. RNA concentration and 28S/18S ratios were determined using an RNA StdSens LabChip® kit and Experion™ Automated Electrophoresis Station (Bio-Rad Laboratories). No-template controls and RNA-RT controls were included in the analysis to verify the absence of contamination. The absence of nonspecific products or primer–dimer artifacts in the samples was confirmed by melting curve analysis and by product visualization using electrophoresis. Primer efficiency of >95 % was confirmed with a standard curve spanning seven orders of magnitude. Data were analyzed using CFX Manager Software (Version 1.5; Bio-Rad Laboratories). Two biological replicates, resulting from two different RNA extractions, were used for analysis, and three technical replicates were analyzed for each biological replicate. The data are presented as mean ± SE

The current hypothesis is that *rolC* suppresses ROS production and *rolB* activates ROS degradation [13]. Consistent with this hypothesis, expression of the NADPH oxidase gene, *RcRBOH3*, is inhibited in *rolC*-transformed *R. cordifolia* cells. RcRBOH3 is a constitutively active oxidase that supports ROS homeostasis under normal conditions. In contrast to *rolC*, the *rolB* gene caused activation of both the inducible and constitutive forms of *R. cordifolia* NADPH oxidase genes, *RcRBOH1* and *RcRBOH3* (Fig. 1a).

Under steady-state conditions, *rolB*-transformed cells of *R. cordifolia* and *Arabidopsis thaliana* maintain a normal redox balance [14]. It is likely that the activation of antioxidant genes is not a consequence of a direct action of *rolB*. We propose that the *rolB*-transformed cells receive an unknown ROS-inducing signal from the RolB protein. The cells attempt to compensate for this effect by cellular compensatory mechanisms, adjusting available antioxidant systems at the optimum time and location [14]. In cases where it is not possible because of excessive *rolB* expression, the cells die by necrosis. RolB possesses a dual function in necrotic processes in plant cells. High doses of RolB cause necrotic cell death, and low doses prevent necrosis. This effect is dependent on the environment: if the RolC protein is expressed in cells along with RolB, the necrotic process is suppressed.

Another feature of *rolB* is that the gene strongly activates expression of the extracellular class III peroxidase genes and enhances total peroxidase activity in transformed cells [61]. This effect is dose-dependent and is observed in all *rolB*-expressing plant species studied. The *rolC* gene does not influence peroxidase expression and activity. Similarly, the peroxidase expression and activity were not changed in pRiA4-transformed calli, suggesting the counterbalancing effect of other T-DNA genes [61]. The biological significance of the observed *rolB*-stimulating effect is not presently clear. Extracellular class III peroxidases, representing an important group of pathogenesis-related proteins of the PR-9 family, are regulators of extracellular H_2O_2 levels and catalysts of the biosynthesis of polyphenolic phytoalexins [1]. RolB may participate in both of these processes through the activation of peroxidases.

2.2 Apoptosis

It has been established that *rolB* not only prevents necrotic cell death but also reduces apoptotic symptoms (symptoms of programmed cell death) in transformed plant cells [30]. Typical views of the control (R) and the *rolB*-expressing cells stained with fluorescent dyes to distinguish apoptotic and necrotic cells are presented in Fig. 2.

The antiapoptotic effect was particularly evident in the cells with induced apoptosis [30]. This is a new finding for plant T-DNA oncogenes. Testing all of the apoptotic features of transformed cells, including the investigation of several model species and the dynamics of the apoptotic process, as well as the evaluation of the specificity of action of other *rol* genes, are future subjects of investigation.

The mechanism by which *rolB* inhibits apoptosis in transformed cell cultures likely includes the up-regulation of antioxidant genes. RolB permanently supports an active antioxidative status of transformed cells and simultaneously maintains their normal redox balance, providing physiological conditions for adaptation to external deleterious stimuli. This peculiarity of RolB may be important for

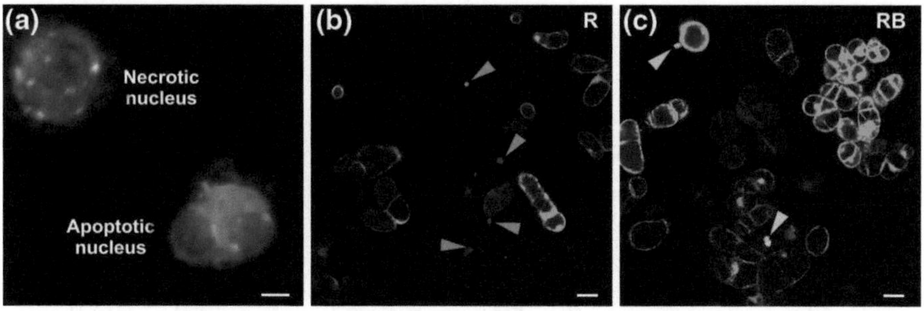

Fig. 2 Expression of *rolB* in *R. cordifolia* cells diminishes apoptosislike symptoms [30]. Cells of the control (R) and *rolB*-transformed (RB) cultures were analyzed by laser confocal microscopy. Propidium iodide and Hoechst 33342 were used to detect collapsed nuclei of necrotic cells and nuclei of apoptotic cells, respectively. **a** A view of nuclei of *R. cordifolia* cells dyed by necrosis (*red*) and apoptosis (*blue* + *red*). In control nontransformed cells (**b**), blue staining of the nuclei (*cyan arrows*) indicates an early phase of apoptosis and violet staining of the nucleus reflects a later phase of apoptosis. Necrotic cells with red-stained nuclei are indicated by *red arrows* (**c**). Scale bars, 2 μm (**a**) and 50 μm (**b**, **c**)

apoptosis prevention, inasmuch as the cellular conditions leading to apoptosis induction are known to be linked with the imbalance of the redox status [51].

2.3 Stress Responses of Transformed Cells

Transformation of plant cells with wildtype *A. rhizogenes* A4 resulted in establishing transformed calli, which tolerate temperature and salt stress [55]. The *rolC* and *rolB* genes expressed alone also induce resistance to salt, low and high temperatures, and excessive light [11, 14]. An example of cell growth measurements under different stress conditions is presented in Fig. 3.

The acquired tolerance of transformed cells to cold treatment is moderate, whereas the tolerance to elevated temperature or salt is higher. The expression of *rolC* completely prevented growth depression caused by high temperatures. *A. rhizogenes* A4-transformed cells showed increased tolerance to elevated temperatures and salt stress. A direct comparison of action of the *rol* genes and pRiA4 is not possible, because cells transformed with pRiA4 or single *rol* genes have different hormonal requirements. Phytohormones are necessary to support the growth of *rolB*- and *rolC*-transformed cells, whereas pRiA4-transformed cells are hormone-autotrophic. In the presence of plant hormones, their growth is substantially inhibited. The interplay between hormones and stress responses of transformed cells is a subject of our current investigation.

A promising example of antistress activity was demonstrated during experiments with acute light stress [11]. The stress was induced in *R. cordifolia* cells by an argon laser, and video files of stressed cells were recorded by confocal microscopy for 16 min ([11]; Fig. 4).

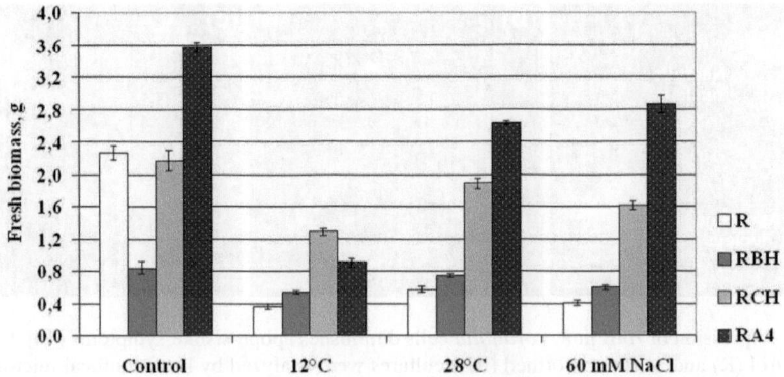

Fig. 3 Resistance of *R. cordifolia* cells to low and high temperatures and salt stress. The data are summarized from earlier publications [11, 14, 55]. Control: normal conditions of cultivation (24 °C). *R* nontransformed control, *RBH* high-*rolB*-expressing cells, *RCH* high-*rolC*-expressing cells, *RA4 A. rhizogenes* A4 transformed cells. The inoculum biomass was 0.2 g fresh weight. The R, RBH, and RCH cultures were grown in the presence of 6-benzylaminopurine (0.5 mg l^{-1}) and α-naphthaleneacetic acid (2.0 mg l^{-1}). The A4 cells were grown in hormone-free medium

During this time, the control cells increased the intracellular ROS level up to fourfold, demonstrating a massive oxidative burst. In contrast, ROS levels in *rolC*-expressing cells were not increased. In these cells, ROS elevations were induced only after a threefold increase in laser intensity (Fig. 4).

Thus, acting as ROS suppressors in transformed cells, the *rolC* and *rolB* genes were able to prevent oxidative damage, which is the main destructive factor of abiotic stresses [65]. This observed phenomenon is somewhat surprising. Many plant pathogens have developed strategies aimed at suppressing plant defense reactions. A well-studied example is *Pseudomonas syringae,* which uses a protein tyrosine phosphatase to combat plant defenses (Sect. 3). *A. rhizogenes*, acting via the *rol* genes, renders cells more tolerant to abiotic stresses. This observation, along with the activation of secondary metabolism and activation of PR-proteins, leads us to suggest that the *rolB* and *rolC* genes activate plant defense reactions.

2.4 T-DNA Oncogenes and Secondary Metabolism

Stimulation of secondary metabolism by the *rolC* and *rolB* genes is a well-known phenomenon [10]. These genes are potential activators of secondary metabolism in transformed cells of the Solanaceae, Araliaceae, Rubiaceae, Vitaceae, and Rosaceae families. Among the activated metabolites were anthraquinones (AQs), alkaloids (tropane, pyridine, and indole groups), stilbenes (resveratrol), isoflavonoids (isoflavones and pterocarpans), and dammarane-type glycosides (ginsenosides). The extent of secondary metabolism activation varies among plant species from 2- to 300-fold depending on the group of secondary metabolites and the plant

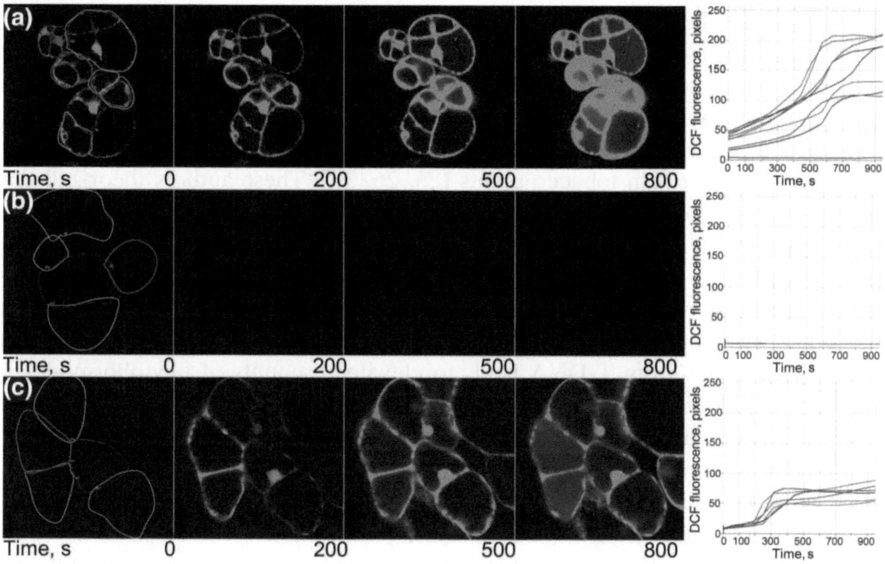

Fig. 4 Expression of *rolC* diminishes light-induced ROS elevation. The figure represents time series of ROS production in control (R) and low-*rolC*-expressing (RCL) cells of *R. cordifolia*. The cells were subjected to illumination from argon laser for 16 min. ROS were visualized using fluorogenic dye H_2DCF-DA (green fluorescence). (**a**) Time series of ROS production in R cells. The intensity of the laser was 5.9 % of maximal. (**b**) ROS in the RCL cells at laser intensity 5.9 %. (**c**) ROS in the RCL cells at laser intensity 17.7 %. Images of single cells were captured and video files of the images were analyzed using LSM 510 Release 3.5 software. *Right panel* represents dynamics of ROS production in individual cells

species [10, 12]. The activation in some cases is remarkably stable over time. In *R. cordifolia* transformed cells, the stimulator effect remains stable over a 12-year period of cultivation.

However, in other plant species, these genes result in more complex outcomes on secondary metabolism. In *Eritrichium sericeum* and *Panax ginseng* transformed cells, the *rolC* gene caused initial inhibition and subsequent stimulation of particular groups of secondary metabolites [9, 15, 35]. It was reported previously that in *Lithospermum erythrorhizon* cells, *rolC* prevented the accumulation of naphthoquinone pigments [9]. Similar to the stimulator effect, the inhibitor effect is also long-lasting, as to date *rolC*-transformed cells of *L. erythrorhizon* cannot produce naphthoquinones after an almost 10-year period of cultivation. The assessment of the signal transduction pathways affected by the gene showed that the *rolC*-gene-mediated signal did not interfere with general plant defense pathways leading to synthesis of phytoalexin-type secondary metabolites. In particular, the Ca^{2+}-dependent NADPH oxidase pathway and the jasmonic acid-mediated pathway were not affected in *rolC*-transformed cells of *R. cordifolia* [7, 8]. ROS production and phytoalexin production are most likely dissociated in *rolC*-transformed cells [11].

In contrast to hairy roots, crown gall tissues have been less studied with respect to the biosynthesis of secondary metabolites. Analysis of polyphenols in C58-*6b*, 35S-AK-*6b*, and dex-AK-*6b* transformed tobacco plants revealed increased levels of chlorogenic acid, caffeoyl putrescine, rutin, and kaempferol-3-rutinoside in transformed plants compared with control plants, an effect that was dependent on the hormonal context in tobacco plants [25, 26, 36]. These authors theorized that *6b* acts as a regulator of the metabolite flux between branches of the biosynthetic pathways rather than as a biosynthetic activator and have put forward an interesting hypothesis that *6b* suppresses polar auxin transport via the increased accumulation of polyphenols. The phenylpropanoid-auxin model [36] suggests that upon infection with *A. tumefaciens*, the expression of auxin and cytokinin biosynthesis genes from T-DNA leads to elevated amounts of phytohormones at the site of infection. The hormones trigger the *6b* gene expression, and the 6b protein activates the biosynthesis of phenylpropanoids, which limits the flow of auxin, creating a high auxin concentration within the infection site. The elevated auxin amounts induce extensive cell proliferation to support the progression of crown galls. Clément and colleagues [19] also found an overaccumulation of chlorogenic acid in *6b*-expressing tissues in tobacco, but they proposed that the effect could be attributed to the defense function of polyphenols against concurrent micro-organisms (this was the first proposition of such a function for Ti oncogenes). According to a provisional sugar accumulation model, sugar accumulation at the site of infection is a leading factor in crown galls progression [18, 19].

Both models are not mutually exclusive, due to the complex action of 6b. The 6b protein seems to possess a dual activity and acts as a transcriptional activator/modulator and mediator of the RNA-silencing machinery (see Sect. 4.1). Redirection of the metabolite flux among the branches of biosynthetic pathways via RNA silencing was recently established [31]. In that study, mir156 targeted an SPL transcription factor to change the MYB-bHLH-WD40 complex formation, thereby redirecting metabolite flux between anthocyanins and flavanols. Because 6b interferes with miRNA (microRNA) pathways and modulates accumulation of diverse miRNAs [64], a regulatory effect on numerous proteins, including those participating in secondary metabolism and sucrose transport, could be expected. The precise mechanism of 6b action on secondary metabolism and sucrose transport remains to be investigated.

The same situation is observed with the RolC and RolB proteins. There is information that these proteins activate secondary metabolism through the up-regulation of key genes of secondary metabolism [35, 37, 54], but the mechanism of activation is unknown. According to the emerging role of plant oncoproteins as modulators of RNA silencing, a mechanism could be proposed by which T-DNA oncoproteins (such as RolB, RolC, or 6b) interfere with RNA silencing to de-repress genes of secondary metabolism normally repressed through miRNA. The activation/de-repression may involve the interaction of the Rol or 6b proteins with transcription factors, which regulate the biosynthesis of secondary metabolites such as MYBs. This is an intriguing new topic in the biology of individual *rol* genes and hairy root cultures. Furthermore, we can now expect the opening of a new horizon in the biotechnology of secondary metabolites. Understanding these mechanisms may contribute to establishing a new generation of plant cell cultures for industrial applications, for which there is an insistent need [15].

3 Tyr Phosphorylation and Microbial Pathogenesis

An interesting analogy emerges between the effects mediated by RolB of *A. rhizogenes* and some type III proteins of *Pseudomonas syringae*. The effector HopAO1 (HopPtoD2) protein of *P. syringae* is injected from bacterial cells into plant cells to promote bacterial growth by the suppression of the innate immunity of host cells. It was shown that HopAO1 possesses protein tyrosine phosphatase activity [3, 22], suppresses induced ROS in plants [3], and suppresses induced programmed cell death [22]. Because RolB possesses a tyrosine phosphatase activity [24], suppresses induced ROS (Sect. 2.1), and suppresses induced programmed cell death (Sect. 2.2), a functional analogy between RolB and HopAO1 is evident.

In Arabidopsis, HopAO1 suppresses several defense responses associated with pathogen-associated molecular pattern (PAMP)-triggered innate immunity [60]. Among the genes down-regulated by HopAO1, genes encoding pathogenesis-related proteins PR-1, PR-2, and PR-5, MYB122, a transcription factor involved in regulation of secondary metabolites, and XET proteins participating in strengthening of the cell wall, were found.

Whereas ROS inhibition by *rol* genes is in line with similar effects of other pathogens, the functionality of stimulation of defense reactions is less clear. It can be hypothesized that activation of defense reactions in transformed plants could provide advantages for *A. rhizogenes* by inhibiting concurrent pathogenic microorganisms. A recent investigation demonstrated that *rolC*-transformed strawberry plants possessed increased tolerance to infection by *Phytophthora cactorum* [39]. An interesting observation was reported about the interaction of *Panax ginseng* calli with the human pathogen *Yersinia pseudotuberculosis* [50], in which the co-cultivation of *Y. pseudotuberculosis* and control (vector-transformed) ginseng calli resulted in the rapid death of both bacterial and plant cells, resembling the HR-syndrome known from plant–microbe interactions. *Y. pseudotuberculosis* caused the death of control ginseng cells but not *rolC*-transformed cells. In this case, *rolC* expression was sufficient to defend plant cells during the fatal interaction. This is also a new area in investigations of T-DNA oncogenes and the question of whether T-DNA oncogenes participate in plant defense reactions against other pathogens requires further examination.

Likewise, the mechanism by which RolB enhances antioxidant defense remains to be determined. Nuclear localization of RolB mediated by 14-3-3 proteins [43] favors the hypothesis that the primary targets of RolB are nuclear proteins. For the HopAO1 tyrosine phosphatase, MAP kinases have been proposed as the targets [3, 22], although this opinion was not supported by other authors [33, 60]. There is a theory that RolB may be involved in the de-phosphorylation of tyrosine residues of multiple nuclear proteins, thereby ensuring pleiotropic effects on plant cells. However, recent data on the 6b protein show that the situation may be more complex.

4 Complicated Functions of Plant Oncogenes: Suppressors of Innate Immunity or Regulators of Cell Adaptability?

4.1 Important New Insights into the Mechanism of 6b Action

Whereas molecular mechanisms determining tumor formation caused by *rol* genes remain unclear, significant progress has been achieved with the plant *6b* oncogene from *A. tumefaciens*. An important observation was made by Leon Otten's group (Strasbourg, France), which revealed significant similarity of the morphological responses caused by *rolC* and another member of the *plast* gene family, the *6b* oncogene [42]. Indeed, both *6b* and *rolC* induced enations with similar morphologies. These enations represent very unusual growth modifications, such as secondary leaf blades arising in a mirrorwise fashion along the veins of the abaxial leaf side [42].

Yasunori Machida's group (Nagoya University, Japan) found that the 6b protein is localized in the nucleus and interacts with nuclear factors Nt SIP1 (*Nicotiana tabacum* 6b-interacting protein 1) and Nt SIP2 [38]. Therefore, the 6b protein is likely a transcriptional co-activator/mediator. The 6b protein could interact not only with Nt SIP1 but also with other nuclear factors in tobacco and Arabidopsis [38]. The level of 6b-interacting proteins in various cell types is different; therefore, the strength of 6b-mediated effects is also different in particular cells and plants. This situation is complicated by the fact that 6b could act as a transcriptional repressor. For example, 6b activated expression of class I *KNOX*, *CUC*, and cell-cycle-related genes [57], and it repressed expression of the *IAA3/SHY2* and *IAA6* genes [58].

Furthermore, this same group established the histone chaperonelike function of 6b [58]. This function implies the disassembly of nucleosomes and the assembly of core histones, thereby initiating transcription by RNA polymerase II and maintaining elongation of RNA. The Nt SIPI protein likely recruits 6b to specific regions of chromatin.

A further important finding was published by Wang and colleagues [64], who showed that 6b is involved in ADP ribosylation by targeting key components in the microRNA processing machinery. The authors studied the crystal structures of the 6b protein and discovered the ADP-ribosylating toxin fold in 6b, which was closely related to that of cholera toxin. Therefore, 6b is presently considered to be an ADP-ribosylating toxin. There are two groups of proteins interacting with 6b: one group targets the repeating acidic Glu residue loop (residues 164–184), and the other group targets other structural elements, such as a well-ordered loop (residues 40–55) [64]. The acidic loop (164–184) serves as a scaffold for protein–protein or protein–nucleic acid interactions, ensuring transcriptional effects of 6b (activation or repression of target genes). The second loop (residues 40–55) serves as an interacting platform for the ADP-ribosylating activity of 6b by binding histone H3, AGO1, and SE. The ectopic expression of a number of genes, such as class I *KNOX*, *CUC*, and cell-cycle-related genes, initiated by *6b* expression likely depends on the general ADP-ribosylating activity of 6b [64]. Interestingly, the *A. rhizogenes* T-DNA gene *orf13* also induced ectopic expression of KNOX class transcription factors and several genes involved in cell cycle control [56]. Notably Orf13, 6b, and RolC form a separate cluster in the phylogenetic tree of the RolB family proteins [34].

Thus, 6b targets a wide range of different host plant proteins in different ways, demonstrating a complex mechanism of action. The combination of transcriptional activator/modulator activity and RNA-silencing activity in a single regulator protein is unique.

4.2 T-DNA Versus Viruses: View on Suppression of Host Defense

Wang and colleagues [64] suggested that T-DNAs can be considered to be functionally related to DNA viruses. They demonstrated a morphological similarity between plants expressing the *6b* oncogene and plants expressing the viral *2b* gene (a known viral gene that suppresses plant immunity). This proposition is interesting. A similar assumption regarding the function of the RolB protein was also reported [43]. These authors hypothesized that in spite of different target proteins, most of the oncoproteins belonging to the RolB family might associate with plant proteins to alter developmental plasticity, as in the case of viral oncoproteins in animal cells [43]. Furthermore, another member of the Plast/RolB proteins, RolA, shares structural similarity with the papilloma virus E2 DNA-binding domain [52]. To shed light on this idea, plant viruses must be introduced into our discussion.

RNA silencing in plants prevents virus accumulation; accordingly, viruses have evolved various strategies to counteract this defense mechanism [2]. As viruses are inducers, suppressors, and targets of the RNA silencing mechanism, there are many ways in which the symptoms in infected plants can be influenced by viral intervention in the microRNA (miRNA) and small interfering RNA (siRNA) pathways. The counterdefense mechanism involves suppressor proteins of silencing that are encoded in the genomes of both RNA and DNA viruses. These proteins most likely evolved independently in different virus groups because they are structurally diverse and there are no common sequence motifs. RNA silencing can be seen as a form of the immune system that operates at the nucleic acid level. However, unlike immune systems mediated by proteins, the specificity of the RNA-silencing component of the immune system is not programmed by the host. Instead, it is determined by the features of the pathogen genome [62]. In RNA interference, a long and perfectly complementary dsRNA is cleaved by Dicer into siRNAs. One siRNA strand is incorporated into the multisubunit RNA-induced silencing complex (RISC) and guides the complex to degrade cellular RNA molecules that are identical in sequence to the siRNA [62]. The argonaute (AGO) proteins are components of the silencing effector complexes that bind to siRNAs. AGO proteins are the "slicer" ribonucleases in RISC. AGO1 is likely a RISC component in *A. thaliana* [2].

In *35S-6b*-transformed plants, Wang et al. [64] found decreased amounts of miR162, miR164, and miR319 transcripts. The partially overlapping miRNA deficiency phenotypes among the 35S-6b plants, se-1 mutant, and ago1-27 mutant prompted the authors to test the possible direct interactions among AK6b, SE, and AGO1. Indeed, the results show that 6b executes its suppressor function by directly targeting SE and AGO1, two key components of the miRNA machinery in Arabidopsis. Therefore, they suggested that both RISC/AGO1 and DCL1-SE-HYL1 are essential protein complexes that participate in virus-induced gene silencing [64]. A functional analogy between *6b* and virus suppressor genes (such as *2b*, *P0*, *HcPro*, and *P19* of positive-strand RNA viruses and *AC2* of DNA viruses) means that plant oncogenes, such as *6b*, should suppress the innate immunity of transformed cells. Is this true for the RolB and RolC oncoproteins? In the following section, we discuss this question.

4.3 Do Plant Oncogenes Cause Adverse Effects on Plant Cells?

This question is not an easy one to answer. Of course, all plant oncogenes disturb normal physiological processes in plants, provoke numerous developmental defects, and cause agricultural losses. The effect of oncogenes, at least *rolC* and *rolB*, on a single transformed cell is less obvious.

Our observations show that the transformation of plant cells by *rolC* caused positive rather than negative effects on the growth of transformed cells and tissues. The main effect is that *rolC*-transformed cells demonstrate a high viability during long-term cultivation [9, 11, 29, 54]. For example, *rolC*-transformed roots of *Panax ginseng*, established almost 20 years ago, continue to grow well up to the present day. In contrast, normal in vitro cultivated roots of this plant demonstrate less viability during a long-term cultivation. *RolC*-low expressing *P. ginseng* calli (2c2 line) is characterized by a remarkably stable and vigorous pattern of growth over a long period (over 12 years) without any selection [29]. Callus cultures of *rolC*-transformed cells of *R. cordifolia* possess a remarkably stable and vigorous growth during a 12-year period of cultivation [54]. High growth and viability have been shown for other *rolC*-transformed cell cultures, such as *Eritrichium sericeum*, *Lithospermum erythrorhizon*, *Maackia amurensis*, and others established more recently in our laboratory. It should be noted that only low and moderate expression of *rolC* induces vigorous cell growth exceeding that of control cells. As mentioned above, this trait of high growth and viability is often accompanied by a high production of secondary metabolites and a resistance to salt and unfavorable temperatures (Sects. 2.3, 2.4). The *rolB* gene acts in a similar fashion, although *rolB* cannot stimulate growth in any doses.

The current hypothesis of the Otten group is that the *rolC* gene plays a role in sucrose transport and metabolism, providing conditions for growth stimulation of transformed organs (hairy roots). The ancient agrobacterial *rolC* gene in tobacco (*trolC*) is functionally active and can play a favorable role in tobacco growth and physiology [42]. This role of *rolC* was also suggested by Nilsson and Olsson [46]. The stimulation of sucrose transport and metabolism may be a part of numerous effects of this oncogene on plant physiology, and this effect cannot be attributed to toxic effects. The Otten group also showed that in crown gall formation, the *6b* gene stimulates a local wound-induced increase in glucose and fructose. The resulting osmolarity change would help to prevent desiccation in wound areas and thereby favor cell division. At the same time, high glucose and fructose levels might contribute to agrobacterial growth, either directly or after incorporation into opines [18].

The similarities and differences in gene silencing by virus- and T-DNA-mediated processes were studied by Dunoyer and colleagues [21]. By investigating the *A. tumefaciens* strain A281 carrying pTiBo542 plasmid and the A208 strain carrying pTiT37 plasmid, with disarmed strain GV3101 as a control, these authors showed that successful infection of plants by *A. tumefaciens* relies on the suppression of the plant immune response mediated by RNA interference. However, the main conclusion was that, in this infection model, T-DNA oncogenes, unlike virus suppressor genes, have only a modest role in tumor progression. RNA silencing against T-DNA genes occurred at early stages of infection but became inhibited upon successful disease development.

Despite that the above-mentioned observations contrast with the hypothesis that T-DNAs are functionally related to DNA viruses [64], we believe that the hypothesis is an interesting one. Where could one find functional similarity between the T-DNA oncogenes and the world of viral genes? We should examine those plant–virus interactions where viruses are not obligate suppressors of plant immunity. It is known that viruses may exploit the host silencing response (see the section, "RNA Silencing to the Benefit of Viruses" [62]). In this case, viruses tolerate rather than suppress RNA silencing, thereby securing a host plant for virus propagation. In this way, viruses reprogram the expression of specific host proteins to establish optimal infection conditions. These conditions include the preservation of the integrity of the plant, restriction of virus overaccumulation, and restriction of the growth of concurrent micro-organisms.

Regarding the physiological role of the *rol* genes, we suggest a similar scenario for the plant–*Agrobacterium* interaction. The idea emerging is that the *rol* genes (and perhaps the *6b* gene) are not suppressors of innate immunity but rather are regulators that provide some functions to plants that can help them to increase adaptability. The boundaries between mutualism and pathogenesis are fluid and mutualistic and pathogenic interactions involve strikingly similar principles [66]. Further investigation of the *rol* genes may help us to understand why closely

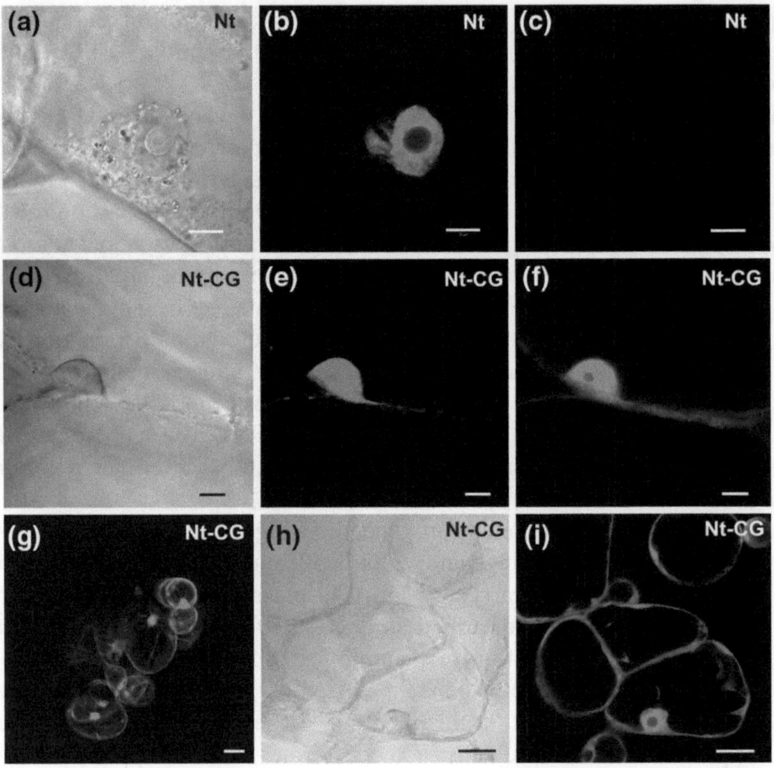

Fig. 5 Nuclear localization of the RolC protein in transformed *Nicotiana tabacum* cells. Tobacco Nt cells were stable and transformed with *A. tumefaciens* GV3101/*35S-rolC-sGFP* as described previously [12] to generate the Nt-CG cell line. Fluorescence of the green fluorescent protein sGFP was excited by an argon laser and monitored by confocal microscopy. Localization of RolC in cells is indicated by green color. DAPI (4′,6-diamidino-2-phenylindole, molecular probes) staining was used to detect nuclei (*blue color*). **a** Light microscopy of a tobacco control cell. **b** The same cell stained with DAPI to detect the nucleus. **c** The same cell under GFP-detecting conditions (excitation: 488 nm; emission filter: 508–525 nm). Note the absence of any fluorescence in this cell. **d** Nt-CG cell, light microscopy. **e** The same cell stained with DAPI. **f** The same cell under GFP-detecting conditions. The GFP signal is localized in the nucleus, but not in the nucleolus. **g–i** Partitioning of the RolC-GFP signal in Nt-CG cells. **g** GFP localization: a general view. Cells under light (**h**) and fluorescent (**i**) microscopy. Note that the RolC-GFP signal is localized in nuclei and cell wall cytoplasm. Scale bars, 50 μm (**g**), 10 μm (**a–f**, **h**, **i**)

related micro-organisms such as *A. rhizogenes*, *A. tumefaciens*, and Rhizobium species have evolved to develop such different interactions with plants as pathogenic or symbiotic.

In summary, the recently discovered functions of the *rolB* and *rolC* genes include (a) modulation of secondary metabolism, (b) modulation of intracellular ROS levels, (c) increased stress resistance of transformed cells, (d) modulation of sucrose transport and metabolism, and (e) inhibition of programmed cell death (at least for *rolB*).

Table 2 A comparison of recently discovered activities of the *rolB*, *rolC*, and *6b* genes

Effects	*rolB*	*rolC*	*6b*	References
ROS suppression	+	+	?	Bulgakov et al. [11, 14] Veremeichik et al. [61]
Increased expression of antioxidant genes	+	–	?	
Stress resistance	+	+	?	
Modulation of secondary metabolism	+	+	+	Gális et al. [25] Kakiuchi et al. [36] Bulgakov [10] Shkryl et al. [54]
Sucrose transport/metabolism	?	+	+	Clément et al. [18, 19] Mohajjel-Shoja et al. [42]
Similar morphology of enations (secondary leaf blades)	–	+	+	Mohajjel-Shoja et al. [42]
Transcriptional activation/modulation	+[a]	+[a]	+	Terakura et al. [57, 58] Shkryl et al. [54] Kiselev et al. [37] Inyushkina et al. [35]
ADP-ribosylating activity	?	?	+	Wang et al. [64]

[a] Measured as the activation of expression of key genes of secondary metabolism

As a consequence of these effects, *A. rhizogenes* seems to be able to establish optimal infection conditions by reorganizing the functioning of host proteins. In many cases, *rolC* and *rolB* demonstrate opposite activity, but their combined action somehow balances unfavorable effects of the individual genes. Related genes *rolA*, *rolD*, and *orf13* may also contribute to this process. We have suggested that the actions of each of the *rol* genes seems to be in tune with the actions of the others, providing physiological conditions for better cell fitness in the face of changing environmental conditions [14]. We believe this is the main effect of the *rol* genes as members of the *RolB* (*plast*) gene family.

It will be interesting to investigate whether *rol* genes have a function analogous to *6b*. According to the phylogenetic analysis [34], *rolC* is more related to *6b* than *rolC* is related to *rolB*. The main drawback is that RolC has been reported to be a cytosolic protein [23]. However, our experiments revealed a nuclear localization of RolC (along with its cytosolic localization) by expression of the *35S-rolC-sGFP* construct in plant cells (Fig. 5). Nuclear localization of RolC was shown both for tobacco and *R. cordifolia* transformed cells. Detailed results on RolC localization will be published elsewhere. The presently known similarities and dissimilarities between the Rol and 6b proteins are presented in Table 2.

5 Conclusions

For some time, studies of the biochemical functions of *A. tumefaciens* and *A. rhizogenes* T-DNA oncogenes expanded in nearly independent ways. This divergence was due to the concept that the pathogens use conceptually different ways to affect plant growth. It was generally accepted that *A. tumefaciens* uses the widely known hormone biosynthesizing genes as the main pathogenic determinants (with some participation of other T-DNA oncogenes) and that *A. rhizogenes* uses the *rol* oncogenes (with some participation of other T-DNA oncogenes). In this way, both pathogens disturb hormone metabolism or hormone perception pathways and attain their aim of successful pathogenesis.

Recent studies described in this work indicate a more complex role for T-DNA oncogenes. The emerging role of the *6b* gene in *A. tumefaciens* pathogenicity is that the gene ensures the transcriptional activator/modulator activity and RNA-silencing activity. The RNA-silencing activity is commonly associated with the virus-induced perturbations of plant defense. Thus, *A. tumefaciens* contains a pathogenic determinant that is potentially connected with plant defense. These studies have revolutionized the investigation of this pathogen, suggesting the existence of not only the hormone-related mechanism of pathogenicity but also the defense-related mechanism of pathogenicity.

At the same time, research on *A. rhizogenes* T-DNA oncogenes shows their role in secondary metabolism, ROS metabolism, and PR-protein expression, thereby indicating the presence of the defense-related mechanism of pathogenicity. On this basis, studies of both pathogens acquire a common direction, which is strengthened by discovery of the common traits of the *rolC*- and *6b*-transformed plants such as altered sugar metabolism and a similarity of morphological defects. Are *A. rhizogenes* oncogenes involved in RNA-silencing and are *A. tumefaciens* oncogenes involved in ROS metabolism and defense reactions? Addressing these questions may be critical for the further investigation of plant–*Agrobacterium* interactions.

Another aspect of these basic findings is that they may contribute to biotechnology and the metabolic engineering of secondary metabolites. There may be two functions of the T-DNA oncogenes. The first is the direct transcriptional activation of key genes of secondary metabolism; the second function is the de-repression of silencing genes. Removal of the persistent negative regulatory controls in cultured plant cells by RNA silencing is a prospective biomimetic approach that can be taken from these studies.

Acknowledgments This work was supported by grants from the Russian Foundation for Basic Research and by the Grant Program "Molecular and Cell Biology" of the Russian Academy of Sciences.

References

1. Almagro L, Gómez Ros LV, Belchi-Navarro S, Bru R, Ros Barceló A, Pedren MA (2009) Class III peroxidases in plant defence reactions. J Exp Bot 60:377–390
2. Baulcombe D (2004) RNA silencing in plants. Nature 431(7006):356–363

3. Bretz JR, Mock NM, Charity JC, Zeyad S, Baker CJ, Hutcheson SW (2003) A translocated protein tyrosine phosphatase of *Pseudomonas syringae* pv. *tomato* DC3000 modulates plant defence response to infection. Mol Microbiol 49:389–400
4. Bonhomme V, Laurain Mattar D, Fliniaux MA (2000) Effects of the *rolC* gene on hairy root: Induction development and tropane alkaloid production by *Atropa belladonna*. J Nat Prod 63:1249–1252
5. Bulgakov VP, Khodakovskaya MV, Labetskaya NV, Chernoded GK, Zhuravlev YN (1998) The impact of plant *rolC* oncogene on ginsenoside production by ginseng hairy root cultures. Phytochemistry 49:1929–1934
6. Bulgakov VP, Tchernoded GK, Mischenko NP, Khodakovskaya MV, Glazunov VP, Zvereva EV, Fedoreyev SA, Zhuravlev YN (2002) Effects of salicylic acid, methyl jasmonate, etephone and cantharidin on anthraquinone production by *Rubia cordifolia* callus cultures transformed with *rolB* and *rolC* genes. J Biotechnol 97:213–221
7. Bulgakov VP, Tchernoded GK, Mischenko NP, Shkryl YN, Glazunov VP, Fedoreyev SA, Zhuravlev YN (2003) Effects of Ca^{2+} channel blockers and protein kinase/phosphatase inhibitors on growth and anthraquinone production in *Rubia cordifolia* cultures transformed by the *rolB* and *rolC* genes. Planta 217:349–355
8. Bulgakov VP, Tchernoded GK, Mischenko NP, Shkryl YN, Fedoreyev SA, Zhuravlev YN (2004) The *rolB* and *rolC* genes activate synthesis of anthraquinones in *Rubia cordifolia* cells by mechanism independent of octadecanoid signaling pathway. Plant Sci 166:1069–1075
9. Bulgakov VP, Veselova MV, Tchernoded GK, Kiselev KV, Fedoreyev SA, Zhuravlev YN (2005) Inhibitory effect of the *Agrobacterium rhizogenes rolC* gene on rabdosiin and rosmarinic acid production in *Eritrichium sericeum* and *Lithospermum erythrorhizon* transformed cell cultures. Planta 221:471–478
10. Bulgakov VP (2008) Functions of *rol* genes in plant secondary metabolism. Biotechnol Adv 26:318–324
11. Bulgakov VP, Aminin DL, Shkryl YN, Gorpenchenko TY, Veremeichik GN, Dmitrenok PS, Zhuravlev YN (2008) Suppression of reactive oxygen species and enhanced stress tolerance in *Rubia cordifolia* cells expressing the *rolC* oncogene. Mol Plant-Microbe Interact 21:1561–1570
12. Bulgakov VP, Shkryl YN, Veremeichik GN (2010) Engineering high yields of secondary metabolites in *Rubia* cell cultures through transformation with *rol* genes. Methods Mol Biol 643:229–242
13. Bulgakov VP, Shkryl YN, Veremeichik GN, Gorpenchenko TY, Inyushkina YV (2011) Application of *Agrobacterium rol* genes in plant biotechnology: a natural phenomenon of secondary metabolism regulation. In: Alvarez M (ed) Genetic transformation. InTech, Rijeka, pp. 261–271. http://www.intechopen.com/articles/show/title/application-of-agrobacterium-rol-genes-in-plant-biotechnology-a-natural-phenomenon-of-secondary-meta
14. Bulgakov VP, Gorpenchenko TY, Veremeichik GN, Shkryl YN, Tchernoded GK, Bulgakov DV, Aminin DL, Zhuravlev YN (2012a) The *rolB* gene suppresses reactive oxygen species in transformed plant cells through the sustained activation of antioxidant defense. Plant Physiol 158:1371–1381
15. Bulgakov VP, Inyushkina YV, Fedoreyev SA (2012b) Rosmarinic acid and its derivatives: biotechnology and applications. Crit Rev Biotechnol 32:203–217
16. Casanova E, Trillas MI, Moysset L, Vainstein A (2005) Influence of *rol* genes in floriculture. Biotechnol Adv 23:3–39
17. Christensen B, Müller R (2009) The use of *Agrobacterium rhizogenes* and its *rol*-genes for quality improvement in ornamentals. Eur J Hort Sci 74:275–287
18. Clément B, Pollmann S, Weiler E, Urbanczyk-Wochniak E, Otten L (2006) The *Agrobacterium vitis* T-6b oncoprotein induces auxin-independent cell expansion in tobacco. Plant J 45:1017–1027

19. Clément B, Perot J, Geoffroy P, Legrand M, Zon J, Otten L (2007) Abnormal accumulation of sugars and phenolics in tobacco roots expressing the *Agrobacterium* T-*6b* oncogene and the role of these compounds in *6b*-induced growth. Mol. Plant-Microbe Interact 20:53–62
20. Doran PM (2009) Application of plant tissue cultures in phytoremediation research: incentives and limitations. Biotechnol Bioeng 103:60–76
21. Dunoyer P, Himber C, Voinnet O (2006) Induction, suppression and requirement of RNA silencing pathways in virulent *Agrobacterium tumefaciens* infections. Nat Genet 38:258–263
22. Espinosa A, Guo M, Tam VC, Fu ZQ, Alfano JR (2003) The *Pseudomonas syringae* type III-secreted protein HopPtoD2 possesses protein tyrosine phosphatase activity and suppresses programmed cell death in plants. Mol Microbiol 49:377–387
23. Estruch JJ, Parets-Soler A, Schmülling T, Spena A (1991) Cytosolic localization in transgenic plants of the rolC peptide from *Agrobacterium rhizogenes*. Plant Mol Biol 17:547–550
24. Filippini F, Rossi R, Marin O, Trovato M, Costantino P, Downey PM, Lo Schiavo F, Terzi M (1996) A plant oncogene as a phosphatase. Nature 379:499–500
25. Gális I, Kakiuchi Y, Simek P, Wabiko H. (2004) Agrobacterium tumefaciens AK-*6b* gene modulates phenolic compound metabolism in tobacco. Phytochemistry 65:169–179
26. Gális I, Simek P, Van Onckelen HA, Kakiuchi Y, Wabiko H (2002) Resistance of transgenic tobacco seedlings expressing the Agrobacterium tumefaciens C58-*6b* gene, to growth-inhibitory levels of cytokinin is associated with elevated IAA levels and activation of phenylpropanoid metabolism. Plant Cell Physiol 43:939–950
27. Gechev TS, Van Breusegem F, Stone JM, Denev I, Laloi C (2006) Reactive oxygen species as signals that modulate plant stress responses and programmed cell death. BioEssays 28:1091–1101
28. Georgiev MI, Agostini E, Ludwig-Müller J, Xu J (2012) Genetically transformed roots: from plant disease to biotechnological resource. Trends Biotechnol 30:528–537
29. Gorpenchenko TY, Kiselev KV, Bulgakov VP, Tchernoded GK, Bragina EA, Khodakovskaya MV, Koren OG, Batygina TB, Zhuravlev YN (2006) The *Agrobacterium rhizogenes rolC*-gene-induced somatic embryogenesis and shoot organogenesis in *Panax ginseng* transformed calluses. Planta 223:457–467
30. Gorpenchenko TY, Aminin DL, Vereshchagina YV, Shkry YN, Veremeichik GN, Tchernoded GK, Bulgakov VP (2012) Can plant oncogenes inhibit programmed cell death? The *rolB* oncogene reduces apoptosis-like symptoms in transformed plant cells. Plant Signal Behav 7(9):1058–1061
31. Gou JY, Felippes FF, Liu CJ, Weigel D, Wang JW (2011) Negative regulation of anthocyanin biosynthesis in Arabidopsis by a miR156-targeted SPL transcription factor. Plant Cell 23:1512–1522
32. Guillon S, Trémouillaux-Guiller J, Pati PK, Rideau M, Gantet P (2006) Hairy root research: recent scenario and exciting prospects. Curr Opin Plant Biol 9:341–346
33. He P, Shan L, Lin N-C, Martin GB, Kemmerling B, Nürnberger T, Sheen J (2006) Specific bacterial suppressors of MAMP signaling upstream of MAPKKK in *Arabidopsis* innate immunity. Cell 125:563–575
34. Helfer A, Pien S, Otten L (2002) Functional diversity and mutational analysis of *Agrobacterium* 6B oncoproteins. Mol Genet Genomics 267:577–586
35. Inyushkina YV, Kiselev KV, Bulgakov VP, Zhuravlev YN (2009) Specific genes of cytochrome P450 monooxygenases are implicated in biosynthesis of caffeic acid metabolites in *rolC*-transgenic culture of *Eritrichium sericeum*. Biochemistry (Moscow) 74:917–924
36. Kakiuchi Y, Gàlis I, Tamogami S, Wabiko H (2006) Reduction of polar auxin transport in tobacco by the tumorigenic *Agrobacterium tumefaciens* AK-*6b* gene. Planta 223:237–247
37. Kiselev KV, Kusaykin MI, Dubrovina AS, Bezverbny DA, Zvyagintseva TN, Bulgakov VP (2006) The *rolC* gene induces expression of a pathogenesis-related β-1,3-glucanase in transformed ginseng cells. Phytochemistry 67:2225–2231

38. Kitakura S, Fujita T, Ueno Y, Terakura S, Wabiko H, Machida Y (2002) The protein encoded by oncogene *6b* from *Agrobacterium tumefaciens* interacts with a nuclear protein of tobacco. Plant Cell 14:451–463
39. Landi L, Capocasa F, Costantini E, Mezzetti B (2009) RolC strawberry plant adaptability, productivity, and tolerance to soil-borne disease and mycorrhizal interactions. Transgenic Res 18:933–942
40. Levesque H, Delepelaire P, Rousé P, Slightom J, Tepfer D (1988) Common evolutionary origin of the central portions of the Ri TL-DNA of *Agrobacterium rhizogenes* and the Ti T-DNAs of *Agrobacterium tumefaciens*. Plant Mol. Biol 11:731–744
41. Lütken H, Clarke JL, Müller R (2012) Genetic engineering and sustainable production of ornamentals: current status and future directions. Plant Cell Rep 31:1141–1157
42. Mohajjel-Shoja H, Clément B, Perot J, Alioua M, Otten L (2011) Biological activity of the *Agrobacterium rhizogenes*-derived *trolC* gene of *Nicotiana tabacum* and its functional relation to other *plast* genes. Mol Plant-Microbe Interact 24:44–53
43. Moriuchi H, Okamoto C, Nishihama R, Yamashita I, Machida Y, Tanaka N (2004) Nuclear localization and interaction of RolB with plant 14-3-3 proteins correlates with induction of adventitious roots by the oncogene *rolB*. Plant J 38:260–275
44. Mullineaux P, Ball L, Escobar C, Karpinska B, Creissen G, Karpinski S (2000) Are diverse signaling pathways integrated in the regulation of *Arabidopsis* antioxidant defence gene expression in response to excess excitation energy? Phil Trans R Soc Lond B 355:1531–1540
45. Nikravesh F, Khavari-Nejad RA, Rahimian H, Fahimi H (2011) Study of antioxidant enzymes activity and isozymes pattern in hairy roots and regenerated plants in *Nicotiana tabacum*. Acta Physiol Plant. doi:10.1007/s11738-011-0838-1
46. Nilsson O, Olsson O (1997) Getting to the root: the role of the *Agrobacterium rhizogenes rol* genes in the formation of hairy roots. Physiol Plant 100:463–473
47. Ono NN, Tian L (2011) The multiplicity of hairy root cultures: prolific possibilities. Plant Sci 180:439–446
48. Otten L, Schmidt J (1998) A T-DNA from the *Agrobacterium tumefaciens* limited-host-range strain AB2/73 contains a single oncogene. Mol Plant-Microbe Interact 11:335–342
49. Palazón J, Cusidó RM, Roig C, Piñol MT (1998) Expression of the *rolC* gene and nicotine production in transgenic roots and their regenerated plants. Plant Cell Rep 17:384–390
50. Persiyanova EV, Kiselev KV, Bulgakov VP, Timchenko NF, Chernoded GK, Zhuravlev YN (2008) Defense response mechanisms of ginseng callus cultures induced by *Yersinia pseudotuberculosis*, a human pathogen. Russian J Plant Physiol 55:748–755
51. de Pinto MC, Tommasi F, De Gara L (2002) Changes in the antioxidant systems as part of the signaling pathway responsible for the programmed cell death activated by nitric oxide and reactive oxygen species in tobacco Bright-Yellow 2 cells. Plant Physiol 130:698–708
52. Rigden D, Carneiro M (1999) A structural model for the RolA protein and its interaction with DNA. Proteins 37:697–708
53. Shadwick FS, Doran PM (2007) Propagation of plant viruses in hairy root cultures: a potential method for in vitro production of epitope vaccines and foreign proteins. Biotechnol Bioeng 96:570–583
54. Shkryl YN, Veremeichik GN, Bulgakov VP, Tchernoded GK, Mischenko NP, Fedoreyev SA, Zhuravlev YN (2008) Individual and combined effects of the *rolA*, *B* and *C* genes on anthraquinone production in *Rubia cordifolia* transformed calli. Biotechnol Bioeng 100:118–125
55. Shkryl YN, Veremeichik GN, Bulgakov VP, Gorpenchenko TY, Aminin DL, Zhuravlev YN (2010) Decreased ROS level and activation of antioxidant gene expression in *Agrobacterium rhizogenes* pRiA4-transformed calli of *Rubia cordifolia*. Planta 232:1023–1032
56. Stieger PA, Meyer AD, Kathmann P, Fründt C, Niederhauser I, Barone M, Kuhlemeier C (2004) The *orf13* T-DNA gene of *Agrobacterium rhizogenes* confers meristematic competence to differentiated cells. Plant Physiol 135:1798–1808
57. Terakura S, Kitakura S, Ishikawa M, Ueno Y, Fujita T, Machida C, Wabiko H, Machida Y (2006) Oncogene *6b* from *Agrobacterium tumefaciens* induces abaxial cell division at late

stages of leaf development and modifies vascular development in petioles. Plant Cell Physiol 47:664–672

58. Terakura S, Ueno Y, Tagami H, Kitakura S, Machida C, Wabiko H, Aiba H, Otten L, Tsukagoshi H, Nakamura K, Machida Y (2007) An oncoprotein from the plant pathogen *Agrobacterium* has histone chaperone-like activity. Plant Cell 19:2855–2865
59. Torres MA, Jones JDG, Dangl JL (2006) Reactive oxygen species signaling in response to pathogens. Plant Physiol 141:373–378
60. Underwood W, Zhang S, He SY (2007) The *Pseudomonas syringae* type III effector tyrosine phosphatase HopAO1 suppresses innate immunity in *Arabidopsis thaliana*. Plant J 52:658–672
61. Veremeichik GN, Shkryl YN, Bulgakov VP, Avramenko TV, Zhuravlev YN (2012) Molecular cloning and characterization of seven class III peroxidases induced by overexpression of the agrobacterial *rolB* gene in *Rubia cordifolia* transgenic callus cultures. Plant Cell Rep 110:183–188
62. Voinnet O (2005) Induction and suppression of RNA silencing: insights from viral infections. Nat Rev Genet 6:206–220
63. Walmsley AM, Doran PM (2012) Foreign protein production using plant cells: opportunities and challenges. Biotechnol Adv 30:385–386
64. Wang M, Soyano T, Machida S, Yang J-Y, Jung C, Chua N-H, Yuan YA (2011) Molecular insights into plant cell proliferation disturbance by *Agrobacterium* protein 6b. Genes Dev 25:64–76
65. Xiong L, Schumaker KS, Zhu J-K (2002) Cell signaling during cold, drought, and salt stress. Plant Cell 14:S165–S183
66. Zamioudis C, Pieterse CMJ (2012) Modulation of host immunity by beneficial microbes. Mol Plant-Microbe Interact 25:139–150

Adv Biochem Eng Biotechnol (2013) 134: 23–54
DOI: 10.1007/10_2013_182

Published Online: 11 April 2013

Perspectives of the Metabolic Engineering of Terpenoid Indole Alkaloids in *Catharanthus roseus* Hairy Roots

Le Zhao, Guy W. Sander and Jacqueline V. Shanks

Abstract This review looks back on how the terpenoid indole alkaloid pathway and the regulatory factors in *Catharanthus roseus* were identified and characterized, and how metabolic engineering, including genetic engineering and metabolic profiling, was conducted based on the gained knowledge. In addition, further examination of the terpenoid indole alkaloid pathway is proposed.

Keywords *Catharanthus roseus* · Genetic engineering · Metabolic engineering · Metabolic profiling · Terpenoid indole alkaloids

Abbreviation

16OMT	16-Hydroxytabersonine-*O*-methyltransferase
AACT	Acetyl-CoA *C*-acetyltransferase
ADH	Acyclic monoterpene primary alcohol dehydrogenase
AS	Anthranilate synthase
CaMV	Cauliflower mosaic virus
C. roseus	*Catharanthus roseus*
CMK	4-Diphosphocytidyl-2-*C*-methyl-D-erythritol kinase
CMS	4-Diphosphocytidyl-2-*C*-methyl-D-erythritol synthase
CPR	Cytochrome P450 reductase
CrBPF	*C. roseus* box P-binding factor
D4H	Desacetoxyvindoline 4-hydroxylase
DAT	Deacetylvindoline 4-*O*-acetyltransferase
DL7H	7-Deoxyloganin 7-hydroxylase
DXR	1-Deoxy-D-xylulose 5-phosphate reductoisomerase
DXS	1-Deoxy-D-xylulose 5-phosphate synthase
E4P	Erythrose 4-phosphate

L. Zhao · J. V. Shanks (✉)
Department of Chemical and Biological Engineering, Iowa State University, Ames IA, 50011, United States
e-mail: jshanks@iastate.edu

G. W. Sander
Department of Chemical Engineering, University of Minnesota Duluth, Duluth MN, 55812, United States

E. coli	*Escherichia coli*
ESI–MS-MS	Electrospray ionization tandem mass spectrometer
ET	Ethylene
G10H	Geraniol 10-hydroxylase
GA	Gibberellic acid
GBF	G-box binding factors
GES	Geraniol synthase
GPPS	Geranyl diphosphate synthase
HDR	1-Hydroxy-2-methyl-2-butenyl 4-diphosphate reductase
HDS	1-Hydroxy-2-methyl-2-butenyl 4-diphosphate synthase
HMGR	3-Hydroxy-3-methylglutaryl CoA reductase
HMGS	3-Hydroxy-3-methylglutaryl CoA synthase
HPLC	High-performance liquid chromatography
IDI	Isopentenyl diphosphate isomerase
IPP	Isopentenyl diphosphate
JA	Jasmonic acid
LAMT	Loganic acid methyl transferase
MALDI-TOF	Matrix assisted laser desorption with time-of-flight detector
MAT	Minovincinine 19-hydroxy-*O*-acetyltransferase
MC	Monoterpene cyclase
MECS	2-*C*-Methyl-D-erythritol 2,4-cyclodiphosphate synthase
MeJA	Methyl jasmonic acid
MEP	2-*C*-Methyl-D-erythritol 4-phosphate
MFA	Metabolic flux analysis
MIAs	Monoterpenoid indole alkaloids
MVA	Mevalonic acid
MVD	Mevalonate 5-diphosphate decarboxylase
MVK	Mevalonate kinase
NMR	Nuclear magnetic resonance
NMT	*N*-methyltransferase
ORCAs	Octadecanoid-responsive *Catharanthus* AP2/ERF domain
PDA	Photo diode array
PEP	Phosphoenolpyruvate
PMK	5-Phosphomevalonate kinase
PRX1	Peroxidase 1
SA	Salicylic acid
SGD	Strictosidine β-D-glucosidase
SLS	Secologanin synthase
STR	Strictosidine synthase
T16H	Tabersonine 16-hydroxylase
T6,7E	Tabersonine 6,7-epoxidase
TDC	Tryptophan decarboxylase

TIAs	Terpenoid indole alkaloids
TSP-MS	Thermospray mass spectrometry
ZCT	Zinc finger *Catharanthus* transcription factor

Contents

1 Introduction

The terpenoid indole alkaloids (TIAs) are a class of over 3,000 natural products containing residues of trytophan or tryptamine (indole part) and terpenoid building blocks derived from secologanin. Some TIAs are proposed to protect plants from pests and pathogens [1]. In addition to the original benefit brought to the plants, these complex natural products and their derivatives are also used as anticancer, antimalarial, and antiarrhythmic agents. The better known and studied plants producing TIAs are *Catharanthus roseus*, *Tabernaemontana divaricata*, and *Rauvolfia serpentina* [2, 3].

The *Catharanthus* alkaloids comprise a large group of about 130 TIAs. Some major products found in *C. roseus* are anhydrovinblastine, vindoline, catharanthine, ajmalicine, and serpentine [4]. Some bisindole alkaloids can disrupt microtubule formation, causing dissolution of mitotic spindles and metaphase arrest in dividing cells, which make them ideal antineoplastic agents. The pharmaceutical *C. roseus* alkaloids are listed in Table 1.

Vinblastine, vincristine, and the many derivatives of vinblastine have been developed as leading pharmaceuticals [6]. The isolation of vinblastine and vincristine from *C. roseus* is costly because of the low levels *in planta* and the diverse

Table 1 Pharmaceutical *C. roseus* alkaloids

Name	Trade name	Treatment
Ajmalicine	Hydroserpan®, Lamuran®	Hypertension, circulatory disorders
Vinblastine	Vincaleukoblastine, Velbe®	Hodgkin's disease, non-Hodgkin lymphomas, testiscarcinomas, breast cancer, and choriocarcinomas
Vincristine	Leurocristine, Oncovin®	Acute leukemia, Hodgkin's disease, non-Hodgkin lymphomas, rhabdomyosarcomas, Wilm's tumors in children, and breast cancer
Anhydrovinblastine		Antineoplastic agent for cervical and lung cancer [5]

alkaloids present. Approximately 500 kg of dried *C. roseus* leaves are required to isolate only 1 g of vinblastine [7]. This motivated a search for alternate approaches, such as chemical synthesis or production in micro-organisms. However, because of the complex compound structures and complicated biosynthesis pathway, these alternatives were not successful. Currently vinblastine, vincristine, and other valuable alkaloids are produced commercially by extraction of the whole plant. To produce these useful anticancer drugs more efficiently, scientists applied plant tissue culture technology [8]. However, the low yield of vinblastine and vincristine by de novo synthesis using callus or suspension cultured cells of *C. roseus* is not promising. A novel process to produce vinblastine developed by Allelix Inc. in Canada is the production of catharanthine by plant cell fermentation and a simple chemical or an enzymatic coupling with commercially available vindoline [9, 10]. This technology was transferred from the Canadian company to Mitsui Petrochemical Industries, Ltd. in Japan for further development, but has not been commercialized to date. In 1990, the yield of catharanthine through fermentation had increased to 230 mg/L per week with high-cell density [11]. Using the partially purified enzymes from the cultured cell of *C. roseus*, anhydrovinblastine was formed with a conversion yield of about 50 % [9]. In the absence of enzymes, the product of the chemical coupling catalyzed by ferric ions was not only anhydrovinblastine, but also vinblastine. The yields of both alkaloids were 52.8 and 12.3 %, respectively. Further development of the chemical catalysis increased the yield of vinblastine from anhydrovinblastine up to 50 % [12].

Several tissue culture strategies, such as cell suspension, embryo, immobilized cell, callus, and shoot cultures are available. One of the most promising tissue culture types is the hairy root culture, resulting from the transfer of T-DNA from the pathogenic soil bacterium, *Agrobacterium rhizogenes*, into the plant genome [13, 14]. The advantages of hairy root compared with whole plant cultures are rapid growth, ease of precursor feeding, inherent genetic stability, and amenability to genetic transformation. Hairy roots grow rapidly in hormone-free liquid media, with doubling times comparable with suspension cells and greater than other tissue cultures [13, 14]. In addition, hairy roots have been cultivated in large-scale bioreactors [15] and compared with suspension cells, hairy roots have inherent genetic stability [14]. However, the major drawbacks of the *C. roseus* hairy root system are the low yields of valuable compounds and the absence of vindoline, a precursor of vinblastine and vincristine.

2 Terpenoid Indole Alkaloid Metabolism

2.1 Overview of TIAs Biosynthesis Pathway

Terpenoid indole alkaloid biosynthesis in *C. roseus* is a complex process with more than 50 biosynthetic events comprising the involved genes, enzymes, regulators, and intra-/intercellular transporters. The broad class of TIAs in *C. roseus* is derived from strictosidine, produced by the coupling of an indole, tryptamine, and a monoterpenoid, secologanin. The tryptamine derives from the shikimate pathway, whereas secologanin is formed from isopentenyl diphosphate (IPP) through the iridoid pathway with a number of steps, some of which are still unknown. IPP itself derives from either the cytosolic mevalonic acid (MVA) pathway or the plastidial 2-*C*-methyl-D-erythritol 4-phosphate (MEP) pathway. Figure 1 shows an overview of the metabolic pathways of the precursors for TIA biosynthesis.

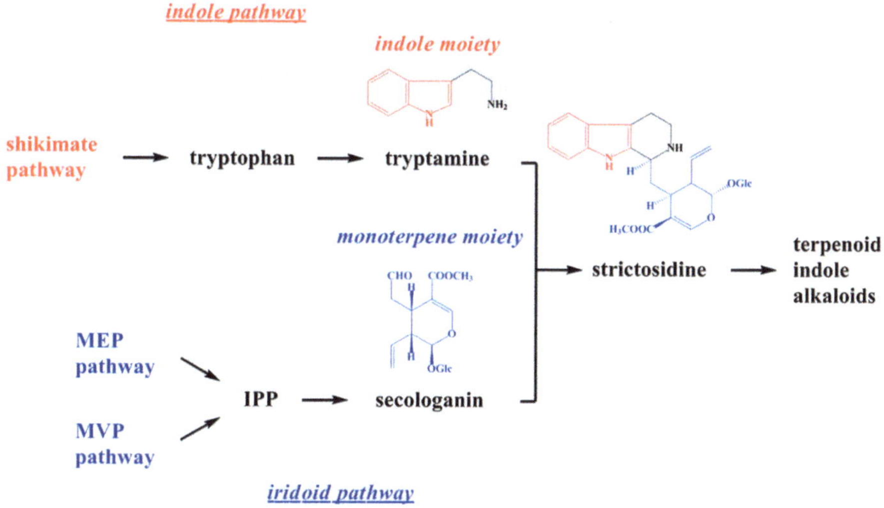

Fig. 1 An overview of the pathways leading to TIA biosynthesis

2.2 Two Metabolic Pathways for IPP Biosynthesis

The MVA pathway was originally considered the sole route leading to IPP; however, the discovery and study of the MEP pathway showed not only its involvement but greater contribution to IPP biosynthesis [16]. The MVA pathway only serves as a minor source of precursors for iridoid biosynthesis, but it contributes through protein prenylation to the regulation of MEP pathway gene expression [17].

The MEP pathway comprises seven steps, of which six have been characterized in *C. roseus*. Despite the impressive progress in the elucidation of the MEP pathway in bacteria and plants (Fig. 2), it is also important to analyze the contribution of the seven enzymes to controlling the flux of intermediates through the pathway. To date, 1-deoxy-D-xylulose 5-phosphate synthase (DXS) is the only enzyme of the MEP pathway shown to have a regulatory role for IPP biosynthesis in all the systems analyzed, including *Arabidopsis*, tomato, and *E. coli* [18]. The role of 1-deoxy-D-xylulose 5-phosphate reductoisomerase (DXR) cannot be generalized for all the systems studied. In *C. roseus* cell cultures, *DXS*, *DXR,* and 2-*C*-methyl-D-erythritol 2,4-cyclodiphosphate synthase (*MECS*) transcript levels had a strong positive correlation with TIA biosynthesis [19, 20]. Other enzymes of the MEP pathway likely also contribute to the regulation of the supply of intermediates for IPP biosynthesis, but this remains to be established.

All the corresponding cDNA expressing enzymes in the MVA pathway in *C. roseus* have been isolated and characterized (Fig. 3), except for 3-hydroxy-3-methylglutaryl CoA synthase (HMGS), which was partially purified in 2004 [4]. 3-hydroxy-3-methylglutaryl CoA reductase (HMGR) is one of the most highly regulated enzymes in animals and plants [21]. The activity of HMGR may be influenced, for example, by post-translational regulation by a protein kinase cascade that phosphorylates and thereby inactivates the enzyme. Allosteric modulation, proteolytic degradation, and the rate of turnover of the corresponding mRNA transcripts may all influence HMGR activity [22].

2.3 Iridoid Pathway

Secologanin is formed from IPP through a number of steps (Fig. 4), some of which remain unknown in *C. roseus*. The enzyme of interest in this pathway is Geraniol 10-hydroxylase (G10H). G10H is a cytochrome P450 monooxygenase thought to play a key regulatory role in TIA biosynthesis [23]. Cytochrome P450 reductase (CPR) transfers electrons from NADPH to G10H. CPR requires the co-factors FMN, FAD, and NADPH. G10H, together with CPR and crude *C. roseus* lipids, can hydroxylate geraniol at its C-10 position [24]. The first evidence that G10H could be a limiting step in the iridoid pathway was the reversible inhibition of G10H activity by catharanthine [25]. Later, G10H activity was found to be

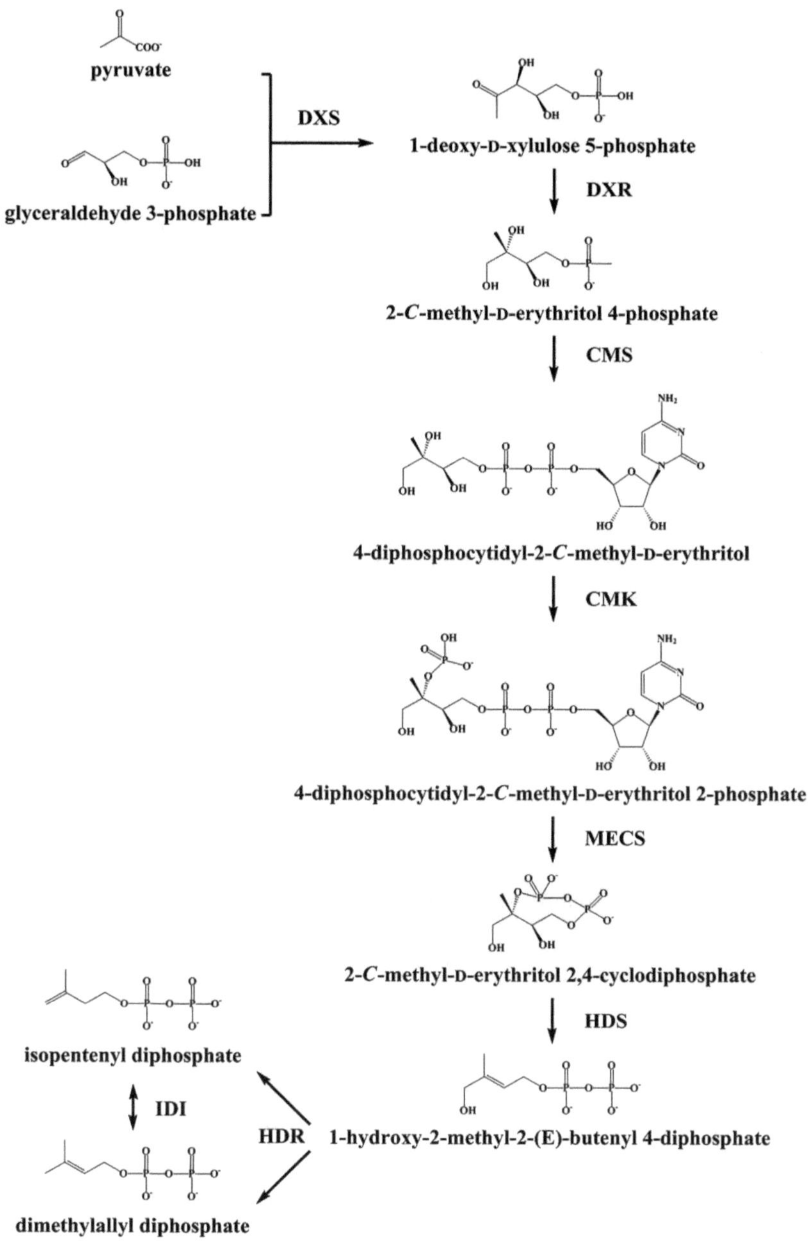

Fig. 2 The 2-*C*-methyl-D-erythritol 4-phosphate (MEP) pathway. *DXS*: 1-deoxyD-xylulose 5-phosphate synthase; *DXR*: 1-deoxyD-xylulose 5-phosphate reductoisomerase; *CMS*: 4-diphosphocytidyl-2-*C*-methyl-D-erythritol synthase; *CMK*: 4-diphosphocytidyl-2-*C*-methyl-D-erythritol kinase; *MECS*: 2-*C*-methyl-D-erythritol 2,4-cyclodiphosphate synthase; *HDS*: 1-hydroxy-2-methyl-2-butenyl 4-diphosphate synthase; *HDR*: 1-hydroxy-2-methyl-2-butenyl 4-diphosphate reductase; *IDI*: isopentenyl diphosphate isomerase

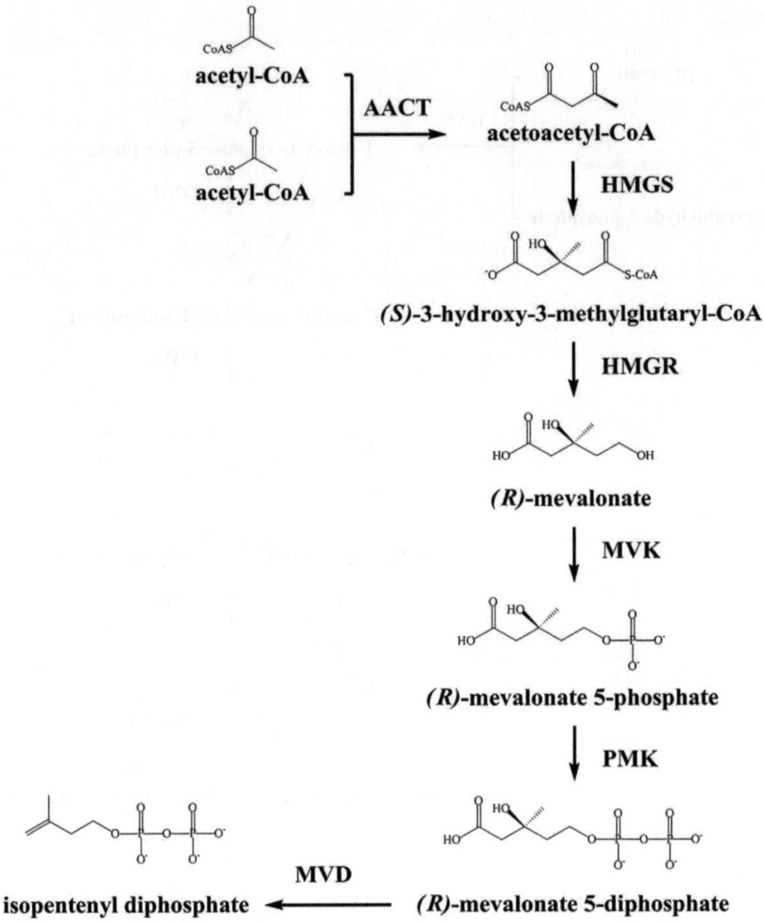

Fig. 3 The mevalonic acid (MVA) pathway. *AACT*: acetyl-CoA *C*-acetyltransferase; *HMGS*: 3-hydroxy-3-methylglutaryl CoA synthase; *HMGR*: 3-hydroxy-3-methylglutaryl CoA reductase; *MVK*: mevalonate kinase; *PMK*: 5-phosphomevalonate kinase; *MVD*: mevalonate 5-diphosphate decarboxylase

positively correlated to TIA accumulation [23]. Loganic acid methyl transferase (LAMT) and secologanin synthase (SLS), which are downstream of G10H, are also proposed as limiting for the synthesis of secologanin [26, 27].

2.4 Indole Pathway

Tryptophan in the indole pathway (Fig. 1) is derived from the shikimate pathway. The shikimate pathway begins with two intermediates of carbohydrate metabolism, phosphoenolpyruvate (PEP) from glycolysis and erythrose 4-phosphate

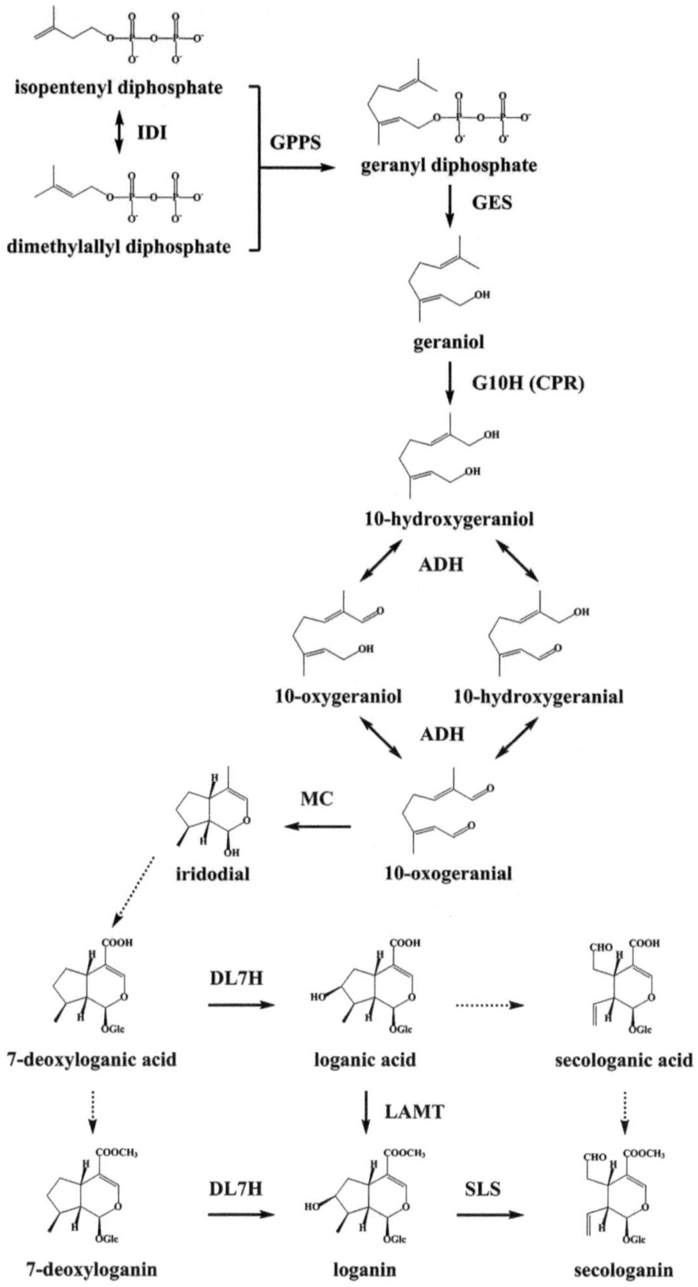

Fig. 4 The iridoid pathway. *IDI*: isopentenyl diphosphate isomerase; *GPPS*: geranyl diphosphate synthase; *GES*: geraniol synthase; *G10H*: geraniol 10-hydroxylase (Cytochrome P450-dependent monooxygenase); *CPR*: Cytochrome P450 reduactase; *ADH*: acyclic monoterpene primary alcohol dehydrogenase (10-hydroxygeraniol oxidoreductase); *MC*: monoterpene cyclase; *DL7H*: 7-deoxyloganin 7-hydroxylase; *LAMT*: loganic acid methyl transferase; *SLS*: secologanin synthase. *Dashed arrows* indicate uncharacterized steps

(E4P) from the pentose phosphate pathway. Tryptophan is converted into tryptamine by tryptophan decarboxylase (TDC). Two highly regulated enzymes in the production of tryptamine are anthranilate synthase (AS), which converts chorismate to anthranilate, and TDC, the last enzyme in this pathway. The AS enzyme is a tetramer composed of two α subunits (ASα) and two β subunits (ASβ), and is feedback-inhibited by tryptophan and tryptamine [28]. The α subunit catalyzes the aromatization of chorismate and binds tryptophan as a feedback inhibitor, and the β subunit donates an amino group from glutamine. TDC is located on the interface between primary and secondary metabolism, and is also feedback inhibited by tryptamine [29].

2.5 Terpenoid Indole Alkaloid Biosynthesis

An overview of the downstream TIA pathway is shown in Fig. 5. Strictosidine is the central intermediate in TIA biosynthesis and is formed by the condensation of secologanin and tryptamine by strictosidine synthase (STR). It is the first committed step in the biosynthesis of TIAs. The expression of the *STR* gene is down-regulated by auxin [30] and up-regulated by fungal elicitation and methyl jasmonic acid (MeJA) [31]. Further studies revealed that hormones and elicitors regulate several transcription factors, which repress or activate the *STR* gene and other genes in the TIA pathway. The transcriptional regulation of the TIA pathway is described in the next section.

Strictosidine β-D-glucosidase (SGD) catalyzes strictosidine and after several spontaneous steps forms 4,21-dehydrogeissoschizine, the precursor for all monoterpenoid indole alkaloids (MIAs; Fig. 6). In one branch, 4,21-dehydrogeissoschizine is converted to form the terminal metabolites serpentine and tetrahydroalstonine. Other branches lead to the important metabolites catharanthine and tabersonine. Eight tabersonine-like MIAs, including lochnericine, all derive from tabersonine and compose a complex pathway where the terminal metabolites are 19-acetylhörhammericine and echitovenine (Fig. 7). The most important pathway starting from tabersonine ends at vindoline after six successive reactions. The MIAs vindoline and catharanthine are condensed to produce the bisindole alkaloids, including vincristine and vinblastine (Fig. 8).

2.6 Regulation of the TIA Pathway

The TIA pathways involve at least 35 intermediates, 30 biosynthetic enzymes, eight regulatory genes, and several cellular compartments [4, 32, 33]. Perhaps not surprisingly, the genetic manipulation of the TIA pathway in *C. roseus* is rarely successful in increasing the final levels of TIA metabolites. Because the TIA pathway is highly regulated, a complete understanding of the regulatory

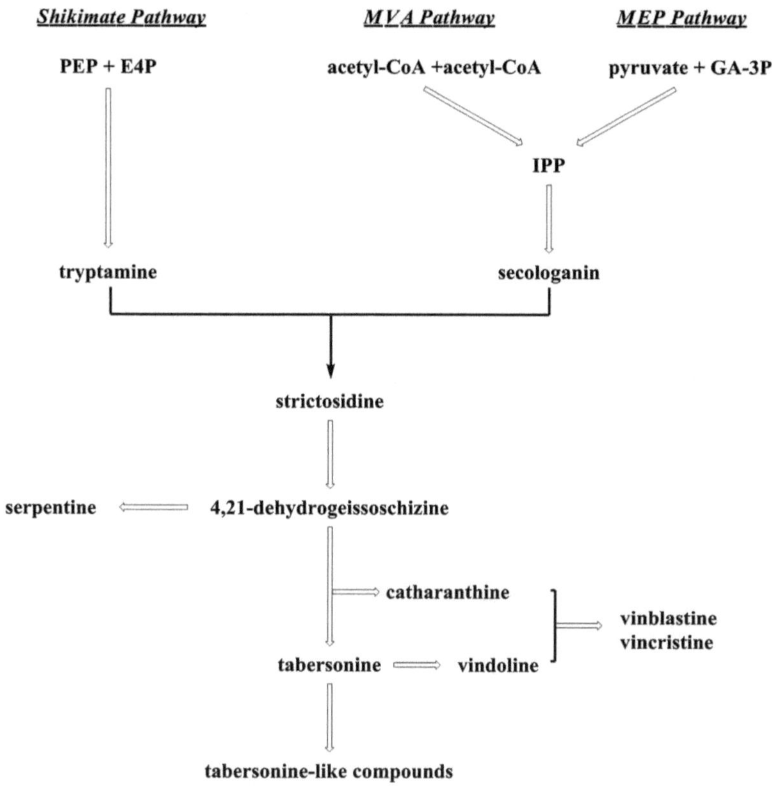

Fig. 5 An overview of the TIA biosynthesis pathway. *PEP*: phosphoenolpyruvate; *E4P*: erythrose 4-phosphate; *IPP*: isopentenyl diphosphate. *Solid arrows* indicate single reactions; hollow arrows indicate multiple steps

mechanisms of the TIA pathway may be a requirement for successful metabolic engineering. Despite the complicated regulation, recent research has shed light on this area.

Extensive research has studied the influence of signaling molecules on TIA biosynthesis. Although the elucidation of the entire regulatory mechanism is not complete, those signaling molecules, such as jasmonate (JA), ethylene, nitrous oxide (NO), and salicylic acid, not only demonstrate involvement in TIA biosynthesis, but also mediate TIA production by either synergistic or antagonistic effects [34].

The most direct regulation of the TIA pathway happens at the transcriptional level. Transcription factors, binding to a specific element and regulating the expression of genes, are a major mechanism controlling the biosynthetic genes of the TIA pathway. Transcription factors are usually regulated by signaling molecules or other elements. During the past few years, much effort was made to identify the transcription factors regulating the TIA pathway and to elucidate which genes are regulated by specific transcription factors.

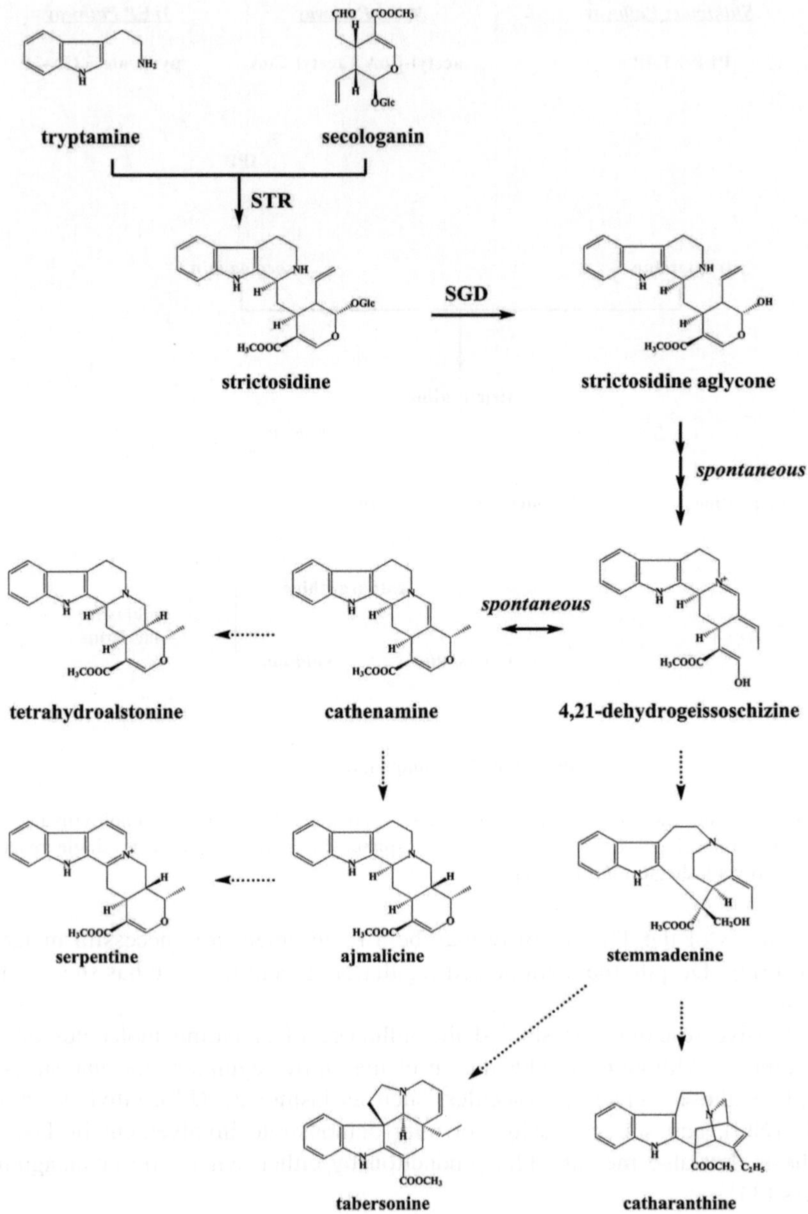

Fig. 6 The main branches of the downstream TIA biosynthesis pathway. *STR*: strictosidine synthase; *SGD*: strictosidine β-D-glucosidase. *Dashed arrows* indicate uncharacterized steps

The most well-known transcription factors are the octadecanoid-responsive *Catharanthus* AP2/ERF domain (ORCAs), including ORCA1, ORCA2, and ORCA3, which are members of the AP2/ERF transcription factor family. ORCA3,

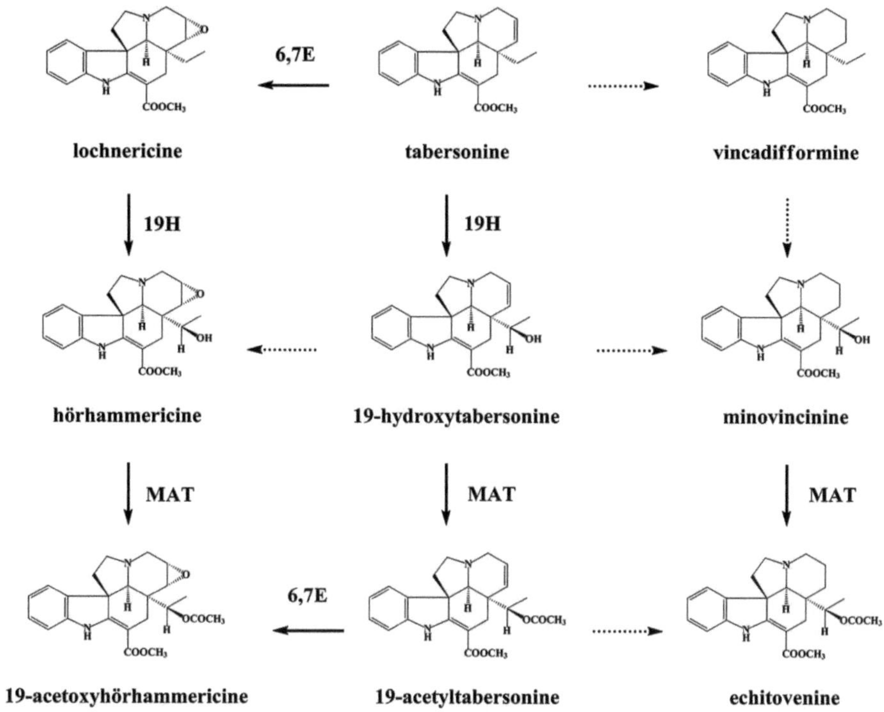

Fig. 7 The biosynthesis of tabersonine-like compounds. 6,7*E*: 6,7-epoxidase; *19H*: 19-hydroxylase; *MAT*: minovincinine 19-hydroxy-*O*-acetyltransferase. *Dashed arrows* indicate uncharacterized steps

for which expression is inducible with JA and fungal elicitation, is the most widely studied. The overexpression of ORCA3 in *C. roseus* hairy roots enhanced the transcript levels of several biosynthetic genes (*AS*, *TDC*, *DXS*, *CPR*, *G10H*, *SLS*, *STR*, *SGD*, and *D4H*) involved in the TIA pathway and consequently increased the accumulation of some TIAs [35]. Previously, overexpression of ORCA3 in *C. roseus* suspension cells also enhanced the expression of *AS*, *TDC*, *DXS*, *CPR*, *STR*, and *D4H* genes[36].

Yeast one-hybrid screening with a *Str* promoter indicated that ORCA2 activates the *Str* promoter and its expression is rapidly inducible with JA and fungal elicitation; in contrast, ORCA1 is expressed constitutively and is not involved in JA- and fungal elicitor-induced *Str* gene expression [31]. Recently, the overexpression of ORCA2 in *C. roseus* hairy roots enhanced the transcript levels of *AS*, *TDC*, *G10H*, *STR*, *D4H*, *T16H*, and *PRX1* genes [37]. In addition to the three ORCAs, another transcription factor, the *C. roseus* box P-binding factor (CrBPF-1) can also bind the *Str* promoter but at a different position. Research suggests that CrBPF-1 may enhance the expression of the *Str* gene when the ORCAs are already bound to the promoter [38]. CrMYC1 [39] and CrMYC2 [40] are basic helix–loop–helix transcription factors. Both of these two transcription factors are induced by JA and

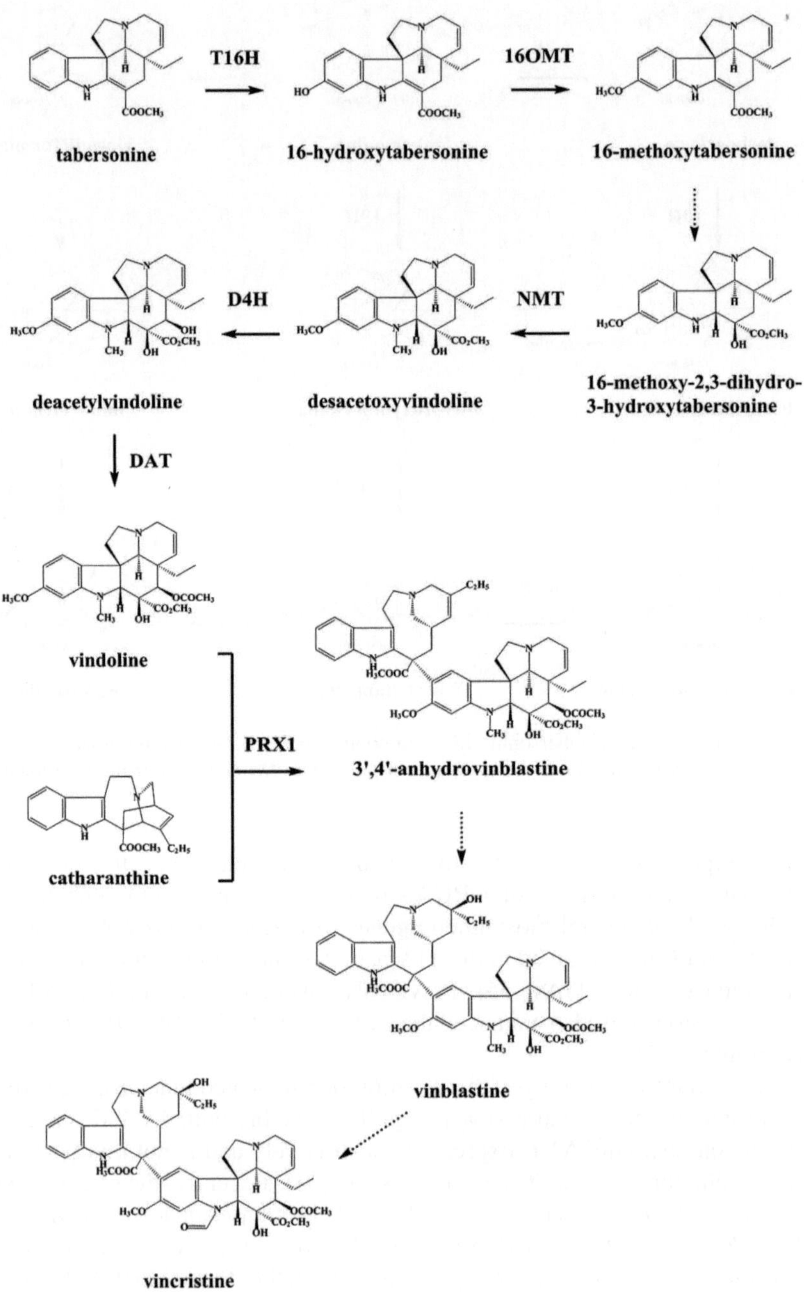

Fig. 8 The biosynthesis of vinblastine and vincristine. *T16H*: tabersonine 16-hydroxylase; *16OMT*: 16-hydroxytabersonine-*O*-methyltransferase; *NMT*: *N*-methyltransferase; *D4H*: desacetoxyvindoline 4-hydroxylase; *DAT*: deacetylvindoline 4-*O*-acetyltransferase; *PRX1*: peroxidase 1. *Dashed arrows* indicate uncharacterized steps

fungal elicitation. The mRNA levels of *CrMYC1* and *Str* genes were enhanced after induction by JA and fungal elicitation, which suggests that CrMYC1 activates the expression of the *Str* gene [39]. CrMYC2 is postulated to act upstream of ORCA2 and ORCA3, activating their transcription [40].

ORCA2, ORCA3, CrBPF-1, CrMYC1, and CrMYC2 act as transcriptional activators for some TIA biosynthetic genes as well as for several transcription repressors for TIA biosynthesis. Three members of the Cys2/His2-type (transcription factor IIIA-type) zinc finger protein family from *C. roseus*, ZCT1, ZCT2, and ZCT3, are activated by ORCA2 and ORCA3, and repress the activity of the promoters of *TDC* and *STR*. In addition, the ZCT proteins can repress the activating activity of the AP2/ERF domain of the ORCAs [41]. In addition to the ZCT proteins, G-box binding factors 1 and 2 (GBF-1 and GBF-2) also act as repressors of *STR* gene expression [42].

CrWRKY1 and CrWRKY2 are jasmonate responsive WRKY transcription factors that activate several genes involved in TIA biosynthesis [43, 44]. Overexpression of *CrWRKY1* in *C. roseus* hairy roots up-regulated *TDC*, as well as the transcriptional repressors *ZCT1*, ZCT2, and *ZCT3*, and down-regulated the transcriptional activators *ORCA2*, *ORCA3*, and *CrMYC2* [43]. In contrast, overexpression of *CrWRKY2* in *C. roseus* hairy roots led to an increase in mRNA transcripts of *TDC*, *NMT*, *DAT*, *MAT*, as well as both specific TIA transcriptional activators (*ORCA2*, *ORCA3*, and *CrWRKY1*) and repressors (*ZCT1* and *ZCT3*) .

The complex interactions of the regulation of TIA biosynthesis is summarized as Fig. 9. Although the isolation of transcription factors regulating the TIA biosynthesis pathway is progressing, further characterization of the regulatory pathways is needed.

3 Metabolic Engineering of TIA Biosynthesis in *C. roseus* Hairy Roots

The metabolic engineering of medicinal plants focuses primarily on increasing the biosynthesis of pharmaceutical metabolites. As more enzymes in the TIA pathway and their corresponding genes have been identified and characterized, as well as the deepening understanding of the whole pathway, it becomes feasible for the metabolic engineering of the TIA pathway to increase the production of valuable pharmaceutical metabolites [45]. Inasmuch as creating transgenic plants of Catharanthus is currently not feasible, tissue and cell culture is the only viable option for an engineered system. The previous metabolic engineering studies of the TIA pathway in *C. roseus* hairy roots are listed in Table 2.

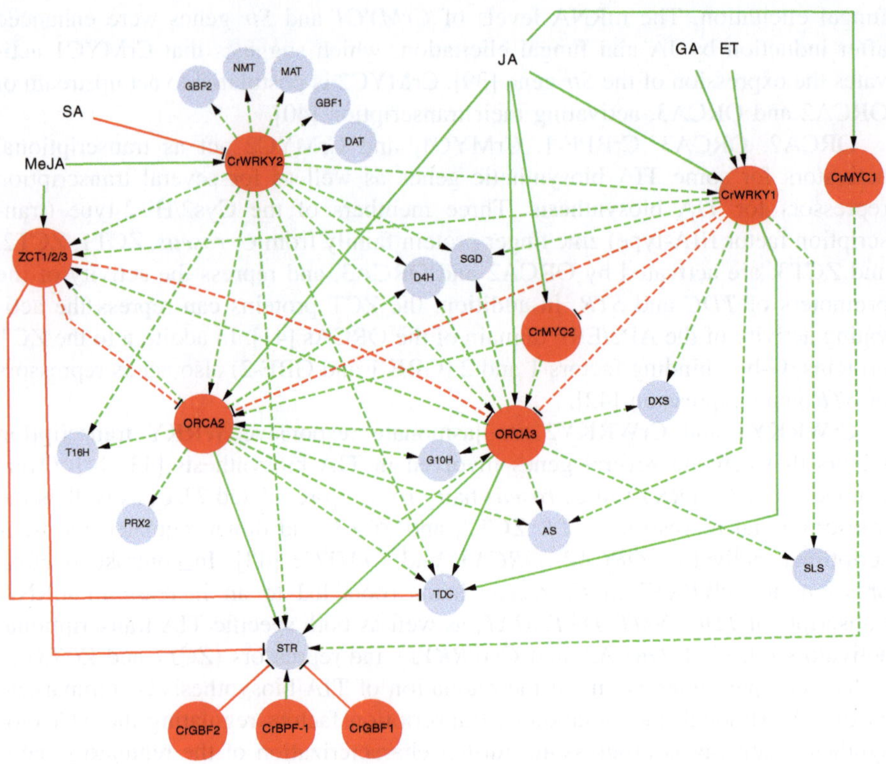

Fig. 9 Regulation map of the TIA biosynthesis pathway. *MeJA*: methyl jasmonic acid; *SA*: salicylic acid; *JA*: jasmonic acid; *ET*: ethylene; *GA*: gibberellic acid. The nodes with *empty* background are signaling molecules; the nodes with *red* background are transcriptional factors; the nodes with *grey* background are genes in the TIA biosynthesis pathway; *dashed lines* represent interactions that may be direct or indirect; *solid lines* indicate potentially direct interactions. *Green lines* with arrows represent transactivation and red lines with bars represent transrepression. *Note* CrWRKY2 overexpressed line has great positive effect on ZCT1, ZCT3 but only slight positive effect on ZCT2; however, in order to make this figure clear, ZCT1, ZCT2, and ZCT3 are lumped

3.1 Metabolic Engineering

AS, the regulatory step of the tryptophan biosynthesis pathway, is under the feedback control of both tryptophan and tryptamine [28]. To test how enhanced tryptophan availability affects the accumulation of indole alkaloids, an *Arabidopsis* feedback-resistant AS α subunit with a glucocorticoid-inducible promoter was overexpressed in *C. roseus* hairy roots. Enzyme assays demonstrated that the heterogeneous α subunit is compatible with the native β subunit and that AS activity is more resistant to tryptophan inhibition in induced than in uninduced cultures. After a 6-day induction, tryptophan and tryptamine yields increased from undetectable levels to 2,500 μg/g dry weight and from 25 to 267 μg/g dry weight,

Table 2 Summary of the generation and metabolic characterization of transgenic *C. roseus* hairy roots lines

Description of Transgenes	Major Observations
Feedback-insensitive *Arabidopsis* ASα [46]	Tryptophan ↑↑↑↑↑, tryptamine↑↑, lochnericine↑
Feedback-insensitive *Arabidopsis* ASα + *C. roseus* Asβ [47]	Tryptophan ↑↑↑↑↑, tryptamine↑↑↑
C. roseus TDC [48]	Serpentine↑
Feedback-insensitive *Arabidopsis* ASα + *C. roseus* TDC [48]	Tryptamine↑↑
Feedback-insensitive *Arabidopsis* ASα + *C. roseus* TDC, *C. roseus* ASβ [47]	Tryptophan ↑↑↑↑↑, tryptamine↑↑↑
Arabidopsis DXS [49]	Serpentine↑, ajmalicine↑, lochnericine↓, hörhammericine↓, tabersonine↓
Arabidopsis DXS + *C. roseus* G10H [50]	Ajmalicine↑, lochnericine↑, tabersonine↑
Arabidopsis DXS + feedback-insensitive *Arabidopsis* ASα [50]	Lochnericine↑, hörhammericine↑, tabersonine↑
C. roseus T16H [49]	Strictosidine↓, catharanthine↓, lochnericine↑, hörhammericine↓, tabersonine↓
C. roseus DAT [51]	Hörhammericine↑
C. roseus ORCA3 [35]	Lochnericine↓, hörhammericine↓, tabersonine↓
C. roseus ORCA2 [37]	Strictosidine↓, secologanin↓, serpentine↑, tabersonine↓↓
C. roseus WRKY1 [43]	Serpentine↑↑, catharanthine↓, tabersonine↓
C. roseus WRKY2 [44]	Serpentine↑, tabersonine↑↑, vindoline↑↑↑

respectively. Although the yield levels of tryptophan and tryptamine increased dramatically, the accumulation levels of most TIAs were not changed, except for lochnericine, which increased 81 % after a 3-day induction [46]. The results indicated that TIA levels are tightly controlled and there are other possible rate-limiting steps in the TIA biosynthesis pathway. Overexpression of TDC with a glucocorticoid-inducible promoter in *C. roseus* hairy roots showed a 48 % increase of TDC activity, a 129 % increase of serpentine yield, and no significant increase in tryptamine levels [48]. In the same study, co-expression of TDC and *Arabidopsis* feedback-resistant ASα resulted in an 87 % increase of TDC activity, sixfold increase of tryptamine for a 3-day late exponential induction, and no effects on TIAs measured. The results suggested that the flux to tryptamine in *C. roseus* hairy roots can be increased, but further manipulation of other biosynthetic steps will be needed to increase the yield of TIAs. To further engineer the indole pathway, an *Arabidopsis* ASβ subunit (ASB1) cDNA was constitutively overexpressed together with the glucocorticoid-inducible overexpression of feedback-resistant ASα and TDC in *C. roseus* hairy roots. Transgenic hairy roots overexpressing the *Arabidopsis* feedback-resistant AS holoenzyme showed significantly greater resistance to feedback inhibition of AS activity than expressing ASα alone, and the AS holoenzyme transgenetic line produced 1.8 times more tryptophan after a 3-day induction. In addition, compared to a fourfold increase of tryptamine levels in the induced ASα line, tryptamine levels in the induced ASα:ASβ line increased up to ninefold. When TDC was expressed together with ASα and ASβ, the induced ASα:ASβ:TDC line produced up to 14-fold higher tryptamine levels than the uninduced lines. Compared with the dramatic increase of tryptophan and tryptamine, the change in alkaloid levels in the induced lines depended on the duration and dosage of induction [47].

Engineering increased flux towards the iridoid pathway, the overexpression of *Arabidopsis* DXS under the control of a glucocorticoid-inducible promoter in hairy root cultures of *C. roseus* led to an increase of loganin, serpentine, and ajmalicine levels and a decrease in tabersonine, lochnericine, and hörhammericine levels [52]. To increase the flux towards downstream metabolites of the TIA pathway, DXS, G10H, and ASα were overexpressed in *C. roseus* hairy roots alone or in combination. The study showed that overexpression of DXS led to an increase in ajmalicine by 67 %, serpentine by 26 %, and lochnericine by 49 % and a decrease in tabersonine by 66 % and hörhammericine by 54 %. Co-overexpression of DXS and G10H resulted in an increase of ajmalicine, lochnericine, and tabersonine by 16, 31, and 13 %, respectively. Similarly, co-overexpression of DXS and ASα showed a significant increase in hörhammericine, lochnericine, and tabersonine by 30, 27, and 34 %, respectively. These results suggested that multiple enzymes may need to be co-overexpressed to increase the accumulation of the downstream metabolites of the TIA pathway [50].

In the *C. roseus* hairy roots system, the pathway from tabersonine to vindoline is blocked. Overexpression of enzymes in the vindoline biosynthesis pathway could be a method to increase the accumulation of valuable metabolites, such as vindoline or even vincristine and vinblastine. Overexpression of tabersonine 16-hydroxylase (T16H), the first enzyme in the vindoline biosynthesis pathway starting from tabersonine, in *C. roseus* hairy roots under the control of a glucocorticoid-inducible promoter was studied. After induction, the level of lochnericine increased, and the levels of strictosidine, catharanthine, hörhammericine, and tabersonine decreased [52]. The last enzyme involved in the vindoline biosynthesis pathway is deacetylvindoline-4-*O*-acetyltransferase (DAT), which converts deacetylvindoline into vindoline. Overexpression of DAT alone with the strong cauliflower mosaic virus (*CaMV*) 35S promoter in *C. roseus* hairy roots did not lead to the accumulation of vindoline. Interestingly, around fourfold higher levels of hörhammericine were observed in the DAT overexpressing line compared with the control line. Enzyme assays showed that the DAT protein could inhibit the enzyme activity of minovincinine-19-*O*-acetyltransferase (MAT), which caused the accumulation of hörhammericine [51].

Transcriptional regulators usually control the expression of multiple genes in a metabolic pathway. It is a reasonable idea metabolically to engineer these transcriptional regulators to enhance the yield of targeted metabolites. In the TIA pathway, several transcriptional regulators have been identified. Among these transcription factors are some that are activators of several TIA biosynthetic genes, such as ORCA2 [53], ORCA3 [36], CrBPF1 [38], CrMYC1 [39], CrMYC2 [40], CrWRKY1 [43], and CrWRKY2 [44], and some that act as repressors, such as the three zinc finger proteins [41], GBF-1, and GBF-2 [42]. The interaction among all the transcription factors is complex, and all of them can be induced by either JA or fungal elicitation. Although many transcriptional regulators have been identified, metabolic engineering with these regulators is still limited. So far, only ORCA2, ORCA3, CrWRKY1, and CrWRKY2 have been overexpressed separately in *C. roseus* hairy roots to study their effect on the regulators, genes, and metabolites in the TIA metabolic pathway [35, 37, 43, 44].

JA treatment and the overexpression of ORCA3 in *C. roseus* hairy roots under a glucocorticoid-inducible promoter were investigated for their effect on TIA metabolite levels and gene expression [35]. The ORCA3 transcripts increased 170-fold after glucocorticoid-induced overexpression and JA elicitation, whereas ORCA3 overexpression without JA elicitation increased ORCA3 transcripts 89-fold, and JA elicitation alone caused a 5-fold increase. Elicitation with JA caused the greatest increase of TIA pathway gene transcripts, including *ASα*, *TDC*, *DXS*, *G10H*, *SLS*, *STR*, and *SGD*, and also caused the greatest increase in tabersonine, lochnericine, hörhammericine, catharanthine, and serpentine concentration. In contrast, induction of ORCA3 overexpression or overexpression plus JA elicitation caused a decrease in tabersonine, lochnericine, and hörhammericine, and did not show a significant effect on transcript levels of TIA pathway genes. The large increase of ORCA3 transcripts did not lead to a significant increase of TIA pathway gene transcripts or TIA metabolites. The observation that the expression of transcriptional repressors ZCT1 and ZCT2 increased along with the increase of the ORCA3 transcripts could possibly explain this.

In addition to ORCA3, the effect of another transcriptional activator, ORCA2, on the TIA pathway has also been studied. ORCA2 was overexpressed under an ethanol-inducible promoter in *C. roseus* hairy roots, and the expression levels of key genes involved in the TIA pathway, as well as the important metabolites, were examined [37]. Compared with the uninduced ORCA2 line, the induced ORCA2 line produced more serpentine, 16-hydroxytabersonine, and 19-hydroxytabersonine and less secologanin, strictosidine, tabersonine, and hörhammericine. Based on the time course analysis of TIA regulatory gene expression, the expression levels of the activator ORCA3 and the repressors ZCT1, ZCT2, and ZCT3 all increased. Although overexpression of the ORCA2 gene could increase the expression of key TIA biosynthetic genes (the side-effect of increasing expression levels of transcriptional repressors produced TIA metabolites), level changes there were transient and limited in magnitude.

A number of independent hairy root lines expressing *CrWRKY1* under the control of the constitutive *CaMV* 35S promoter were generated by Suttipanta et al. [43]. The expression levels of the *CrWRKY1* and the *TDC* genes increased dramatically, 22-fold higher and 7- to 9-fold, respectively, in the CrWRKY1-overexpressing lines compared with the empty vector control lines. Although to a lesser extent than *TDC*, the expression of *AS*, *DXS*, *SLS*, and *SGD* also increased by an average of 1.2-, 0.7-, 2.3-, and 2.4-fold, respectively, compared with the control. For the transcription factors, the expression of ORCA2, ORCA3, and CrMYC2 were suppressed by an average of 90, 80, and 60 %, respectively, and the expression of ZCT1, ZCT2, and ZCT3 increased by an average of 1.4-, 0.3-, and 3.2-fold. To further clarify the role of CrWRKY1 in the regulation of *TDC*, gene-silencing technology was applied and opposite expression patterns were observed for each gene compared with the CrMRKY1 overexpression lines. In the metabolic profiling experiments, the significant increase of serpentine and the reduced production of catharanthine and tabersonine in CrWRKY1 hairy roots indicated that overexpression of CrWRKY1 appears preferentially to activate the serpentine pathway at the expense of the catharanthine and tabersonine pathways.

Suttipanta also did a similar analysis for CrWRKY2-overexpressed hairy root lines as the CrWRKY1 lines [44]. A comparision of the qRT-PCR results of the CrWRKY2 lines under the control of the constitutive *CaMV* 35S promoter with the empty vector control lines revealed that the mRNA transcripts of *TDC*, *STR*, *SGD*, *D4H*, *T16H*, *NMT*, *DAT*, and *MAT* increased. For the transcription factors, ORCA2 expression and ORCA3 expression were significantly enhanced by 4.6- to 16.2-fold and 14.0- to 17.30-fold, respectively. CrWRKY1 (2.5- to 7.4-fold) was moderately up-regulated, and the transcriptional repressors, ZCT1 (4.75- to 7.92-fold), ZCT2 (0.87- to 0.92-fold), ZCT3 (2.47- to 5.15-fold), GBF1 (0.67- to 0.96-fold), and GBF2 (0.74- to 0.82-fold), were also increased to different extents. In the metabolic profiling analysis, the increased accumulation of serpentine (1.6-fold) and tabersonine (5.1- to 6.0-fold) was observed in the transgenic hairy root lines compared with the empty vector control hairy root lines.

All previous research showed that overexpressing the transcriptional activators increased the accumulation of some TIA metabolite levels, while also increasing the transcriptional activators and repressors in the TIA biosynthesis pathway. This indicates that the interaction network of transcription factors is complex. In order to further increase TIA accumulation, the down-regulation of one or more transcriptional repressors in combination with the up-regulation of transcriptional activators may be a promising direction for achieving a sustained increase of TIA gene expression and metabolite levels.

3.2 Metabolic Profiling

Metabolic profiling, the quantification of small biomolecules or metabolic fluxes, provides a direct view of the physiological status of the cell at a certain time point and under specific circumstances [54]. Metabolic profiling is important for understanding cellular regulation, identification of bottlenecks in biosynthetic pathways, and even gaining more knowledge of biological processes.

C. roseus is a well-studied medicinal plant and the two well-known anticancer pharmaceuticals, vincristine and vinblastine, are just 2 of 130 alkaloids produced. However, only a small number ($\sim$11) of the alkaloids are frequently analyzed and even fewer ($\sim$8) are available commercially. In the last decade, the improvement in analytical equipment and protocols provides the ability to detect, separate, identify, and quantify intermediates and end metabolites. The development of large spectral analytical techniques allows the identification of more *C. roseus* secondary metabolites. On the other hand, with the recent perspectives of metabolic engineering to enhance the yield of secondary metabolites [48, 55, 56], it is crucial to be able to identify and quantify the fluxes controlling the secondary metabolic pathways. Therefore, the quantification of the TIAs and of the indole and iridoid precursors enables the process of TIA metabolic engineering.

High-performance liquid chromatography (HPLC) coupled with different detectors, such as the photo diode array (PDA), fluorescence detector, electrospray ionization tandem mass spectrometer (ESI–MS-MS), nuclear magnetic resonance (NMR), and circular dichroism (CD) [57], is efficient for precise identification and/or quantification of alkaloids. Because of the aromatic rings in TIA structures, UV absorbance is effective in TIA detection. The presence of conjugated π-electrons, especially in the aromatic rings, provides intense fluorescent activity, allowing for the possibility of fluorescence detection. Alkaloids are also easily ionized by electrospray in positive mode, therefore ESI (+) MS or MS/MS detection is recommended. Among all the detectors, UV and fluorescence are efficient and common for the quantification of TIAs, whereas the other detectors, such as MS or NMR, are usually used for identifying or more precisely quantifying TIA metabolites.

For UV absorbance detection, 218 nm is the optimal wavelength for maximizing the absorbance level for most of the alkaloids if only a single wavelength is available. However, a higher precision of detection is possible by using 254 nm to quantify strictosidine, ajmalicine, serpentine, catharanthine, and vindoline and 329 nm to quantify tabersonine-like compounds, with signal-to-noise ratios of 4 and 60 times higher than at 218 nm, respectively. In addition to UV detection, fluorescence detection is applied because it is more sensitive, having a lower limit of detection at the micromolar level. The optimal excitation and emission wavelengths in fluorescence detection [58] and the absorption UV wavelengths in UV detection [52] for several *C. roseus* TIAs and precursors are listed in Table 3. Recent developments in the separation column, PDA detectors, as well as fluorescence detectors have contributed to decreasing the limits of detection significantly. However, the common bottleneck of these two detectors for metabolite quantification is the presence of other compounds at the peaks of interest (co-elution) [59].

The major limitation in developing effective separation and identification methods relies on the lack of standards for most of the *C. roseus* secondary metabolites. In order to solve this problem, a semi-preparative (semi-prep) scale HPLC method was developed to purify standards of interest. An example of an analytical method scaled up to obtain tabersonine, lochnericine, and hörhammericine standards from *C. roseus* hairy root tissue is given by Sander [52]. To collect the fractions, the same HPLC system as the analytical method was used with the addition of a fraction collector. Data extracted at 329 nm were used to identify

Table 3 Fluorescence and UV properties of *C. roseus* TIAs and precursors

Alkaloid	Fluorescence [58] (Max Ex/Em in nm)	Absorption UV [52] (nm)
Tryptophan	270/370	218, 278
Tryptamine	270/370	218, 278
Geraniol		199
10-Hydroxygeraniol		199
Loganin		223, 241
Secologanin		222, 240
Strictosidine	280/354	222
Ajmalicine	270/370	225, 279
Serpentine	350/450	248, 305, 363
Catharanthine	290/363	225, 281
Tabersonine		224, 298, 323
Lochnericine		225, 298, 327
Hörhammericine		225, 298, 325
Vindoline	307/357	216, 251, 304
Vinblastine	297/364	216, 266
Vincristine		220, 254, 295

tabersonine, hörhammericine, and lochnericine using retention time standards [60, 61]. The UV chromatograms of the analytical and semi-preparative scale HPLC methods at 329 nm are shown in Fig. 10. The time scales of the analytical chromatogram and the semi-prep chromatogram are different due to the difference in column diameter and flow rate.

Another similar semi-prep method, developed from the Tikhomiroff and Jolicoeur [62] analytical method, separated serpentine, catharanthine, lochnericine, tabersonine, and several unknown compounds. The fractions achieved from a semi-prep method may not be pure because it is possible that several compounds co-elute [63]. The fractions can be separated again under further modified HPLC conditions. The collected fractions could then be further analyzed through MS–MS or ^{1}H-NMR detection.

Widely utilized UV and fluorescence detection applications focus on identifying and quantifying compounds with standards. MS and NMR techniques identify uncharacterized compounds and quantify more precisely. Utilization of MS detection is well established for the *C. roseus* hairy root extracts where all kinds of MS instruments, such as TSP-MS (Thermospray Mass Spectrometry), ESI–MS (Electrospray Ionization Mass Spectrometry), or MALDI-TOF (Matrix-assisted Laser Desorption with Time-of-Flight detector), enhance the identification and quantification of low concentration compounds and of nonpure peaks with a higher precision than UV and fluorescence spectra. In addition, tandem mass or ^{1}H-NMR spectrometry identifies fragment patterns for metabolites at the same molecular weight. For instance, 13 alkaloids from *C. roseus* have the same molecular weight of 352.432 g/mol. The MS/MS fragmentation patterns of some TIAs are listed in Table 4 [64]. HPLC-NMR has been used for the screening of

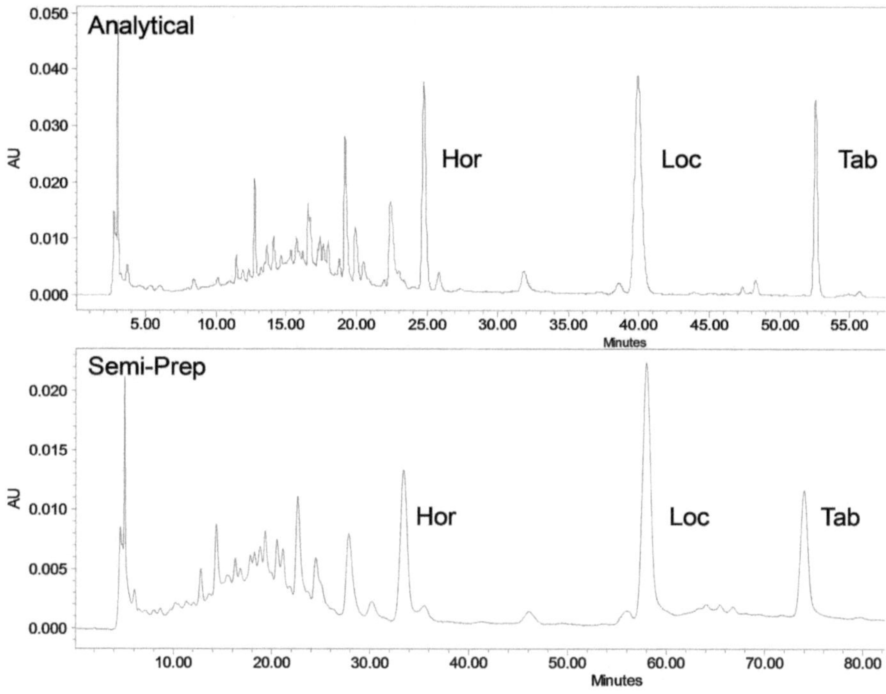

Fig. 10 Chromatograms of analytical and semi-preparative scale HPLC methods at 329 nm

plant extracts [57, 65]. However, this technique has never been used for *C. roseus* hairy root extracts.

Metabolic flux analysis (MFA), the quantification of the intracellular fluxes in a metabolic network, is an important cornerstone for metabolic engineering and systems biology. MFA in one organism determines the phenotype [66] or cell physiology [67]. Comparison of fluxes between phenotypes is of the utmost

Table 4 MS/MS fragmentation patterns of *C. roseus* TIAs

Alkaloid	MS/MS fragmentation pattern [64]	
	Precursor ion $[M + H]^+$ (m/z)	Main fragment ion (m/z)
Ajmalicine	353	222; 210; 178; 144; 117
Serpentine	349	317; 289; 263
Catharanthine	337	174; 144
19*S*-Vindolinine	337	320; 308
Vindolinine	337	320; 308
Vindolidine	427	409; 367; 158; 143
Tabersonine	337	323; 305; 274
Vindoline	457	439; 397; 188
Vinblastine	811	793; 751; 733; 680; 649; 542; 522; 355; 337
Vincristine	825	807; 765; 747; 723; 705; 687

importance to elucidate metabolic regulation [68] and the impact of genetic manipulation to the phenotypes [69], as well as the identification of bottlenecks in product formation, which provides the basis for further metabolic engineering [70].

In MFA of central carbon metabolism, the intracellular fluxes are calculated by using a stoichiometric model for the major intracellular reactions and applying mole balances around intracellular metabolites. A set of measured extracellular fluxes and the growth of the organism are used as inputs to the calculations. In addition, this method requires the assumption of biological steady-state, or homeostasis. One of the advantages of this method is that it does not require the intracellular metabolite concentrations and the enzyme kinetics of the system. However, as the metabolic pathways in plant cell and tissue cultures are far less characterized than microbial or mammalian cultures, this method is currently not feasible for many plant systems. Extension of these MFA models to include secondary metabolism is problematic, inasmuch as the flux to secondary metabolites is frequently less than a few percent of the total carbon flux. Thereby, error in measurement of secondary metabolites would increase dramatically compared to the primary metabolites. Consequently, flux is difficult to calculate using the traditional stoichiometric model. Therefore, the basis of the flux estimations is made directly from intracellular alkaloid or intermediate concentration measurements and focuses specifically on a single class of secondary metabolites and pathways.

Morgan et al. [71] and Sander [49] performed this type of MFA on the TIA pathway in the *C. roseus* hairy roots. The MFA model was based on the following equation,

$$r_i = \frac{d(C_i X)}{dt}$$

where r_i is the flux of a specific alkaloid (mmol alkaloid/h), C_i is the alkaloid specific yield (mmol/g DW), X is the biomass (g DW), and t is the time (h). The values for the variables C_i and X are taken directly from alkaloid and biomass measurements. Metabolic flux maps were derived through a software platform after the input of the model and experimental parameters [49, 71].

4 Conclusion and Outlook

In this review, we have summarized the TIA pathway, transcriptional regulatory networks, and the metabolic engineering of the TIA pathway based on the current progress made in *C. roseus*. Although extensive efforts have been conducted in the last few years, more needs to be done in order to further increase TIA production to a commercialized application.

4.1 Further Elucidation of the TIA Biosynthesis Pathway

Complete elucidation of the TIA biosynthesis pathway is of significant importance for the metabolic engineering of this pathway in *C. roseus* plants or hairy roots. Alternatively, in the future it may be possible to construct large portions of the TIA pathway in microbes for the production of valuable TIAs. Although more and more enzymes and their encoding genes involved in the TIA pathway have been characterized, many steps are still obscure. The vindoline biosynthesis pathway starting from tabersonine involves six enzymatic steps, among which five steps have been characterized; another step converting 16-methoxytabersonine to 16-methoxyl-2,3-dihydro-3-hydroxyltabersonine remains to be elucidated. If the missing genes involved in this pathway can be identified, it would be possible to produce vindoline or even vincristine and vinblastine in *C. roseus* hairy roots. Besides the vindoline biosynthesis pathway, the enzymes involved in many other TIA pathways have not been identified. With the elucidation of these TIA pathways, overexpressing the pathway genes and blocking the competitive pathway by RNAi can be combined to increase the yields of many valuable TIAs.

Identification of the unknown genes involved in the TIA pathway in *C. roseus* is not easy compared to other model plants such as *Arabidopsis*, because the genome sequence is not available yet and there is no mutant library of *C. roseus*. However, recent efforts have been exerted to discover new genes involved in this pathway. By sequencing the cDNA library generated from the RNA of *C. roseus* leaf epidermis cells, the LAMT gene was identified and characterized [72]. Indeed, the identification of an unknown gene requires the integration of several types of information, such as the metabolite profiles of *C. roseus* plants under different growth conditions or different elicitor treatment, the tissue-specific accumulation of certain metabolites, and the related gene expression and regulation patterns. The basic assumption is that the differential accumulation of the TIA metabolites is caused by the differential expression of pathway genes. So, identifying the differentially expressed genes provides a candidate list for the target unknown pathway genes. Many approaches can be used to identify differentially expressed genes, such as cDNA microarray and the newest RNA sequencing technology. The candidate list can be narrowed down by comparing the gene expression in many different samples (RNA extracted from different tissues, plants under different treatment). When a handful of candidate genes are identified, their accurate expression patterns can be analyzed and more detailed bioinformatic analysis can be performed to choose the most promising candidate for further biochemical characterization. By combining all these techniques, the unknown step may eventually be elucidated.

As the cost of genome sequencing decreases, the genome sequence of *C. roseus* may be reported in the near future. Then, the identification of unknown genes may become easier. Eventually, all steps of the TIA pathway may be elucidated.

4.2 Further Elucidation of Regulatory Mechanisms

Limiting improvements to the TIA pathway in *C. roseus* is the unclear correspondence of the DNA, RNA, protein, and metabolite levels to the regulatory mechanisms. Phytohormones are involved in many aspects of plant growth and development, including TIA's biosynthesis pathway, however, how to regulate the molecular mechanism of phytohormones in the TIA pathway is unclear. Forward and reverse genetic strategies should be used to identify important related molecular components in plant hormone perception, signaling, and transport involved in the TIA pathway. For example, it was demonstrated that auxin down-regulated the transcription level of the STR gene in *C. roseus* cell culture [73]. Research on the auxin signaling pathway [74, 75] would help to explore the relationship between plant hormones and TIAs. Calcium signaling is also involved in the regulation of TIA accumulation [76]. The identification of the calcium signaling components involved in the TIA pathway would also be important in the regulation. The use of some new related techniques, such as reverse-genetics approaches, gene-to-metabolite networks, metabolomics analysis, expressed sequence tags, and transcriptome, proteome, and metabolomic analysis, would deepen our understanding of the interrelationships between the TIA pathway and regulatory factors [77].

4.3 Further Development of Metabolic Engineering

In this review, the metabolic engineering of the overexpression some of the key enzymes in the TIA pathway and the two transcriptional activators, ORCA2 and ORCA3, did not increase the accumulation of TIA metabolites significantly. The transcript and metabolic profiling experiments drew a conclusion that down-regulation of transcriptional repressors in combination with up-regulation of transcriptional activators may be a promising direction for achieving higher and more sustained increases in TIA gene expression and metabolite levels.

In the previous metabolic engineering attempts of the TIA pathway in *C. roseus* hairy roots, the maximal number of manipulated genes is three [47]. Multigene transformation in plants is very difficult due to the technical limitations of existing methods. In the past several decades, various plasmids suitable for the cloning, transfer, and expression of foreign genes in plant cells have been developed [78]. The multigene overexpression could be achieved either by constructing a vector containing multiple genes or transferring multiple single-gene vectors in one host cell. For example, a constructed vector with 10 foreign DNA fragments was transferred into the rice genome in 2002 [79], and a mixture of 12 different single-gene plasmids was transformed into soybean embryogenic suspension cultures [80]. If we could manipulate multiple important genes in the TIA pathway as well as the regulatory factors using those techniques in the *C. roseus* hairy root system, it is possible to compel the hairy roots to produce vinblastine and vincristine.

4.4 Further Development of Metabolic Profiling Technology

The full potential of metabolic profiling in plant sciences has not been achieved yet. Several methodological limitations hamper the development of metabolic profiling and need to be addressed in the near future. The major obstacle of MS-based metabolic profiling is that the rate of the analyte identification is currently low because of the lack of structural information obtainable by any kind of mass spectrometer. Although NMR is commonly used to identify small organic molecules, due to the low sensitivity the technique requires time and expense to enrich the natural products for NMR structure analysis. Recently, a new technology, high-performance liquid chromatography combined with solid-phase extraction and NMR spectroscopy (HPLC-SPE-NMR), has been applied to rapidly identify secondary metabolites in some plants. For example, the structures of a total of 15 compounds, some major as well as some minor constituents, of *Harpagophytum procumbens* (Devil's claw) extracts, have been determined by using this technique [81]. Although there are limits to the sensitivity of the HPLC-SPE-NMR technique, it still provides a rapid strategy to identify the metabolites in a complex plant extract. In conclusion, NMR, MS-based technologies coupled with chromatographic separation assays present a chance to develop metabolic profiling to the comprehensive analysis of complex biological systems, such as the TIA metabolites in *C. roseus* hairy roots.

Acknowledgments We thank Al Fu for helping to draw the regulation map of the TIA biosynthesis pathway, Fig. 9. This material is based upon work supported by the National Science Foundation under Award No. CBET-1066879. Any opinions, findings, and conclusions or recommendations expressed in this material are those of the author(s) and do not necessarily reflect the views of the National Science Foundation.

References

1. Luijendijk TJC, van der Meijden E, Verpoorte R (1996) Involvement of strictosidine as a defensive chemical in *Catharanthus roseus*. J Chem Ecol 22(8):1355–1366. doi: 10.1007/bf02027718
2. Waterman P (1998) Chemical taxonomy of alkaloids. In: Roberts MF, Wink M (eds) Alkaloids: biochemistry, ecology, and medicinal applications. Plenum Press, New York, pp 87–107
3. Cordell G (1999) The monoterpene alkaloids. In: Cordell G (ed) The alkaloids: chemistry and physiology. Academic Press, San Diego, pp 261–376
4. van der Heijden R, Jacobs DI, Snoeijer W, Hallard D, Verpoorte R (2004) The *Catharanthus* alkaloids: pharmacognosy and biotechnology. Curr Med Chem 11(5):607–628
5. Schmidt B, Kutney J, Mayer L (1998) Anhydrovinblastine for the treatment of cervical and lung cancer. Patent EP0969839 A1, filed Mar 4, 1998, issued Jan 12, 2000. http://www.google.com/patents/EP0969839A1?cl=en
6. Vantellingen O, Sips JHM, Beijnen JH, Bult A, Nooijen WJ (1992) Pharmacology, bioanalysis and pharmacokinetics of the vinca alkaloids and semisynthetic derivatives. Anticancer Res 12(5):1699–1715

7. Noble RL (1990) The discovery of the vinca alkaloids—chemotherapeutic agents against cancer. Biochem Cell Biol 68(12):1344–1351
8. Carew DP (1966) Growth of callus tissue of *Catharanthus roseus* in suspension culture. J Pharm Sci 55(10):1153–1154. doi:10.1002/jps.2600551041
9. Endo T, Goodbody A, Vukovic J, Misawa M (1988) Enzymes from *Catharanthus roseus* cell suspension cultures that couple vindoline and catharanthine to form 3',4'-anhydrovinblastine. Phytochemistry 27(7):2147–2149
10. Goodbody A, Endo T, Vukovic J, Misawa M (1988) The coupling of catharanthine and vindoline to form 3',4'-anhydrovinblastine by haemoproteins and haemin. Planta Med 54(3):210–214. doi:10.1055/s-2006-962406
11. Fujita Y, Hara Y, Morimoto T, Misawa M (1990) Semisynthetic production of vinblastine involving cell cultures of *Catharanthus roseus* and chemical reaction. In: Nijkamp HJJ, van der Plas LHW, van Aartrijk J (eds) Progress in plant cellular and molecular biology : proceedings of the VIIth International Congress on Plant Tissue and Cell Culture. Kluwer Academic Publishers, Amsterdam, pp 738–743
12. Bede J, Dicosmo F (1992) Enzymatic synthesis of alpha-3',4'-anhydrovinblastine: optimization and immobilization. Planta Med 58(7 SUPPL. 1):A576
13. Flores HE, Hoy MW, Pickard JJ (1987) Secondary metabolites from root cultures. Trends Biotechnol 5(3):64–69. doi:10.1016/s0167-7799(87)80013-6
14. Hamill JD, Parr AJ, Rhodes MJC, Robins RJ, Walton NJ (1987) New routes to plant secondary products. Nat Biotechnol 5(8):800–804. doi:10.1038/nbt0887-800
15. Jung G, Tepfer D (1987) Use of genetic transformation by the Ri T-DNA of *Agrobacterium rhizogenes* to stimulate biomass and tropane alkaloid production in *Atropa belladonna* and *Calystegia sepium* roots grown in vitro. Plant Sci 50(2):145–151. doi: 10.1016/0168-9452(87)90151-8
16. Contin A, van der Heijden R, Lefeber AWM, Verpoorte R (1998) The iridoid glucoside secologanin is derived from the novel triose phosphate/pyruvate pathway in a *Catharanthus roseus* cell culture. FEBS Lett 434(3):413–416. doi:10.1016/s0014-5793(98)01022-9
17. Courdavault V, Burlat V, St-Pierre B, Giglioli-Guivarc'h N (2005) Characterisation of CaaX-prenyltransferases in *Catharanthus roseus*: relationships with the expression of genes involved in the early stages of monoterpenoid biosynthetic pathway. Plant Sci 168(4):1097–1107. doi:10.1016/j.plantsci.2004.12.010
18. Rodríguez-Concepción M, Boronat A, Rodríguez-Concepción M, Boronat A (2002) Elucidation of the methylerythritol phosphate pathway for isoprenoid biosynthesis in bacteria and plastids. A metabolic milestone achieved through genomics. Plant Physiol 130(3):1079–1089. doi:10.1104/pp.007138
19. Chahed K, Oudin A, Guivarc'h N, Hamdi S, Chenieux JC, Rideau M, Clastre M (2000) 1-Deoxy-D-xylulose 5-phosphate synthase from periwinkle: cDNA identification and induced gene expression in terpenoid indole alkaloid-producing cells. Plant Physiol Bioch 38(7–8):559–566. doi:10.1016/s0981-9428(00)00781-6
20. Veau B, Courtois M, Oudin A, Chenieux JC, Rideau M, Clastre M (2000) Cloning and expression of cDNAs encoding two enzymes of the MEP pathway in *Catharanthus roseus*. Biochim Biophys Acta 1517(1):159–163
21. Friesen JA, Rodwell VW (2004) The 3-hydroxy-3-methylglutaryl coenzyme-A (HMG-CoA) reductases. Genome Biol 5(11):248. doi:10.1186/gb-2004-5-11-248
22. Ness GC, Chambers CM (2000) Feedback and hormonal regulation of hepatic 3-hydroxy-3-methylglutaryl coenzyme A reductase: the concept of cholesterol buffering capacity. Proc Soc Exp Biol Med 224(1):8–19
23. Suttipanta N, Pattanaik S, Gunjan S, Xie CH, Littleton J, Yuan L (2007) Promoter analysis of the *Catharanthus roseus* geraniol 10-hydroxylase gene involved in terpenoid indole alkaloid biosynthesis. Biochim Biophys Acta 1769(2):139–148. doi:10.1016/j.bbaexp.2007.01.006
24. Canto-Canche BB, Loyola-Vargas VM (2000) Non-coordinated response of cytochrome P450-dependent geraniol 10-hydroxylase and NADPH: Cyt C (P-450) reductase in *Catharanthus roseus* hairy roots under different conditions. Phyton-Int J Exp Bot 66:183–190

25. McFarlane J, Madyastha KM, Coscia CJ (1975) Regulation of secondary metabolism in higher plants. Effect of alkaloids on a cytochrome P-450 dependent monooxygenase. Biochem Biophys Res Commun 66(4):1263–1269. doi:10.1016/0006-291X(75)90495-7
26. Arvy MP, Imbault N, Naudascher F, Thiersault M, Doireau P (1994) 2,4-D and alkaloid accumulation in periwinkle cell suspensions. Biochimie 76(5):410–416
27. Contin A, Rvd Heijden, Verpoorte R (1999) Accumulation of loganin and secologanin in vacuoles from suspension cultured *Catharanthus roseus* cells. Plant Sci 147(2):177–183. doi: 10.1016/S0168-9452(99)00115-6
28. Li J, Last RL (1996) The *Arabidopsis thaliana trp5* mutant has a feedback-resistant anthranilate synthase and elevated soluble tryptophan. Plant Physiol 110(1):51–59
29. Noe W, Mollenschott C, Berlin J (1984) Tryptophan decarboxylase from *Catharanthus roseus* cell suspension cultures purification, molecular and kinetic data of the homogenous protein. Plant Mol Biol 3(5):281–288. doi:10.1007/bf00017782
30. Pasquali G, Goddijn OJ, de Waal A, Verpoorte R, Schilperoort RA, Hoge JH, Memelink J (1992) Coordinated regulation of two indole alkaloid biosynthetic genes from *Catharanthus roseus* by auxin and elicitors. Plant Mol Biol 18(6):1121–1131
31. Menke FL, Champion A, Kijne JW, Memelink J (1999) A novel jasmonate—and elicitor-responsive element in the periwinkle secondary metabolite biosynthetic gene *Str* interacts with a jasmonate—and elicitor-inducible AP2-domain transcription factor, ORCA2. EMBO J 18(16):4455–4463. doi:10.1093/emboj/18.16.4455
32. Facchini PJ, De Luca V (2008) Opium poppy and Madagascar periwinkle: model non-model systems to investigate alkaloid biosynthesis in plants. Plant J 54(4):763–784. doi: 10.1111/j.1365-313X.2008.03438.x
33. Burlat V, Oudin A, Courtois M, Rideau M, St-Pierre B (2004) Co-expression of three MEP pathway genes and geraniol 10-hydroxylase in internal phloem parenchyma of Catharanthus roseus implicates multicellular translocation of intermediates during the biosynthesis of monoterpene indole alkaloids and isoprenoid-derived primary metabolites. Plant J 38(1):131–141. doi:10.1111/j.1365-313X.2004.02030.x
34. El-Sayed M, Verpoorte R (2004) Growth, metabolic profiling and enzymes activities of *Catharanthus roseus* seedlings treated with plant growth regulators. Plant Growth Regul 44(1):53–58. doi:10.1007/s10725-004-2604-5
35. Peebles CA, Hughes EH, Shanks JV, San KY (2009) Transcriptional response of the terpenoid indole alkaloid pathway to the overexpression of ORCA3 along with jasmonic acid elicitation of *Catharanthus roseus* hairy roots over time. Metab Eng 11(2):76–86. doi: 10.1016/j.ymben.2008.09.002
36. van der Fits L, Memelink J, van der Fits L, Memelink J (2000) ORCA3, a jasmonate-responsive transcriptional regulator of plant primary and secondary metabolism. Sci Signal 289(5477):295. doi:10.1126/science.289.5477.295
37. Li CY, Leopold AL, Sander GW, Shanks JV, Zhao L, Gibson SI (2013) The *ORCA2* transcription factor plays a key role in regulation of the terpenoid indole alkaloid pathway. (manuscript submitted)
38. van der Fits L, Zhang H, Menke FL, Deneka M, Memelink J (2000) A *Catharanthus roseus* BPF-1 homologue interacts with an elicitor-responsive region of the secondary metabolite biosynthetic gene *Str* and is induced by elicitor via a JA-independent signal transduction pathway. Plant Mol Biol 44(5):675–685
39. Chatel G, Montiel G, Pre M, Memelink J, Thiersault M, Saint-Pierre B, Doireau P, Gantet P (2003) *CrMYC1*, a *Catharanthus roseus* elicitor- and jasmonate-responsive bHLH transcription factor that binds the G-box element of the strictosidine synthase gene promoter. J Exp Bot 54(392):2587–2588. doi:10.1093/jxb/erg275
40. Zhang HT, Hedhili S, Montiel G, Zhang YX, Chatel G, Pre M, Gantet P, Memelink J (2011) The basic helix-loop-helix transcription factor CrMYC2 controls the jasmonate-responsive expression of the ORCA genes that regulate alkaloid biosynthesis in *Catharanthus roseus*. Plant J 67(1):61–71. doi:10.1111/j.1365-313X.2011.04575.x

41. Pauw B, Hilliou FA, Martin VS, Chatel G, de Wolf CJ, Champion A, Pre M, van Duijn B, Kijne JW, van der Fits L, Memelink J (2004) Zinc finger proteins act as transcriptional repressors of alkaloid biosynthesis genes in *Catharanthus roseus*. J Biol Chem 279(51):52940–52948. doi:10.1074/jbc.M404391200
42. Sibéril Y, Benhamron S, Memelink J, Giglioli-Guivarc'h N, Thiersault M, Boisson B, Doireau P, Gantet P (2001) *Catharanthus roseus* G-box binding factors 1 and 2 act as repressors of strictosidine synthase gene expression in cell cultures. Plant Mol Biol 45(4):477–488
43. Suttipanta N, Pattanaik S, Kulshrestha M, Patra B, Singh SK, Yuan L (2011) The transcription factor CrWRKY1 positively regulates the terpenoid indole alkaloid biosynthesis in *Catharanthus roseus*. Plant Physiol 157(4):2081–2093. doi:10.1104/pp.111.181834
44. Suttipanta N (2011) Characterization of *G10H* promoter and isolation of WRKY transcription factors involved in *Catharanthus* terpenoid indole alkaloid biosynthesis pathway. University of Kentucky, Lexington
45. Hughes EH, Shanks JV (2002) Metabolic engineering of plants for alkaloid production. Metab Eng 4(1):41–48. doi:10.1006/mben.2001.0205
46. Hughes EH, Hong SB, Gibson SI, Shanks JV, San KY (2004) Expression of a feedback-resistant anthranilate synthase in *Catharanthus roseus* hairy roots provides evidence for tight regulation of terpenoid indole alkaloid levels. Biotechnol Bioeng 86(6):718–727. doi:10.1002/bit.20081
47. Hong SB, Peebles CA, Shanks JV, San KY, Gibson SI (2006) Expression of the *Arabidopsis* feedback-insensitive anthranilate synthase holoenzyme and tryptophan decarboxylase genes in *Catharanthus roseus* hairy roots. J Biotechnol 122. doi:10.1016/j.jbiotec.2005.08.008 (The Netherlands)
48. Hughes EH, Hong SB, Gibson SI, Shanks JV, San KY (2004) Metabolic engineering of the indole pathway in *Catharanthus roseus* hairy roots and increased accumulation of tryptamine and serpentine. Metab Eng 6(4):268–276. doi:10.1016/j.ymben.2004.03.002
49. Sander G (2009) Quantitative analysis of metabolic pathways in *Catharanthus roseus* hairy roots metabolically engineered for terpenoid indole alkaloid overproduction. Iowa State University, Ames, Iowa
50. Peebles CAM, Sander GW, Hughes EH, Peacock R, Shanks JV, San K-Y (2011) The expression of 1-deoxy-d-xylulosesynthase and geraniol-10-hydroxylase or anthranilate synthase increases terpenoid indole alkaloid accumulation in *Catharanthus roseus* hairy roots. Metab Eng 13(2):234–240. doi:10.1016/j.ymben.2010.11.005
51. Magnotta M, Murata J, Chen J, Luca VD (2007) Expression of deacetylvindoline-4-O-acetyltransferase in *Catharanthus roseus* hairy roots. Phytochemistry 68(14):1922–1931. doi:10.1016/j.phytochem.2007.04.037
52. Sander G (2009) Quantitative analysis of metabolic pathways in *Catharanthus roseus* hairy roots metabolically engineered for terpenoid indole alkaloid overproduction. Iowa State University, Ames, Iowa
53. Menke FLH, Champion A, Kijne JW, Memelink J (1999) A novel jasmonate—and elicitor-responsive element in the periwinkle secondary metabolite biosynthetic gene *Str* interacts with a jasmonate—and elicitor-inducible AP2-domain transcription factor, ORCA2. EMBO J 18(16):4455–4463. doi:10.1093/emboj/18.16.4455
54. Nielsen J, Oliver S (2005) The next wave in metabolome analysis. Trends Biotechnol 23(11):544–546. doi:10.1016/j.tibtech.2005.08.005
55. Verpoorte R, van der Heijden R, ten Hoopen HJG, Memelink J (1999) Metabolic engineering of plant secondary metabolite pathways for the production of fine chemicals. Biotechnol Lett 21(6):467–479. doi:10.1023/a:1005502632053
56. Capell T, Christou P (2004) Progress in plant metabolic engineering. Curr Opin Biotechnol 15(2):148–154. doi:10.1016/j.copbio.2004.01.009
57. Bringmann G, Wohlfarth M, Rischer H, Schlauer J, Brun R (2002) Extract screening by HPLC coupled to MS–MS, NMR, and CD: a dimeric and three monomeric naphthylisoquinoline alkaloids from *Ancistrocladus griffithii*. Phytochemistry 61(2):195–204

58. Hisiger S, Jolicoeur M (2007) Analysis of *Catharanthus roseus* alkaloids by HPLC. Phytochem Rev 6(2–3):207–234
59. Chu I-H, Bodnar JA, Bowman RN, White EL (2006) Determination of vincristine and vinblastine in *Catharanthus roseus* plants by high performance liquid chromatography/ electrospray ionization mass spectrometry. J Liq Chromatogr Relat Technol 20(8):1159–1174 (April 1997)
60. Peebles CA, Hong SB, Gibson SI, Shanks JV, San KY (2005) Transient effects of overexpressing anthranilate synthase alpha and beta subunits in *Catharanthus roseus* hairy roots. Biotechnol Prog 21(5):1572–1576. doi:10.1021/bp050210l
61. Morgan JA, Barney CS, Penn AH, Shanks JV (2000) Effects of buffered media upon growth and alkaloid production of *Catharanthus roseus* hairy roots. Appl Microbiol Biotechnol 53(3):262–265
62. Tikhomiroff C, Jolicoeur M (2002) Screening of *Catharanthus roseus* secondary metabolites by high-performance liquid chromatography. J Chromatogr A 955(1):87–93
63. Mans DR, da Rocha AB, Schwartsmann G (2000) Anti-cancer drug discovery and development in Brazil: targeted plant collection as a rational strategy to acquire candidate anti-cancer compounds. Oncologist 5(3):185–198
64. Zhou H, Tai Y, Sun C, Pan Y (2005) Rapid identification of vinca alkaloids by direct-injection electrospray ionisation tandem mass spectrometry and confirmation by high-performance liquid chromatography-mass spectrometry. Phytochem Anal 16(5):328–333
65. Bringmann G, Wohlfarth M, Heubes M (2000) Observation of exchangeable protons by high-performance liquid chromatography-nuclear magnetic resonance spectroscopy and high-performance liquid chromatography-electrospray ionization mass spectrometry: a useful tool for the hyphenated analysis of natural products. J Chromatogr A 904(2):243–249
66. Ratcliffe RG, Shachar-Hill Y (2005) Revealing metabolic phenotypes in plants: inputs from NMR analysis. Biol Rev Camb Philos Soc 80(1):27–43
67. Sauer U (2004) High-throughput phenomics: experimental methods for mapping fluxomes. Curr Opin Biotechnol 15(1):58–63. doi:10.1016/j.copbio.2003.11.001
68. Sweetlove LJ, Fernie AR (2005) Regulation of metabolic networks: understanding metabolic complexity in the systems biology era. New Phytol 168(1):9–24. doi: 10.1111/j.1469-8137.2005.01513.x
69. Ratcliffe RG, Shachar-Hill Y (2006) Measuring multiple fluxes through plant metabolic networks. Plant J 45(4):490–511. doi:10.1111/j.1365-313X.2005.02649.x
70. Sweetlove LJ, Last RL, Fernie AR (2003) Predictive metabolic engineering: a goal for systems biology. Plant Physiol 132(2):420–425
71. Morgan JA, Shanks JV (2002) Quantification of metabolic flux in plant secondary metabolism by a biogenetic organizational approach. Metab Eng 4(3):257–262
72. Murata J, Roepke J, Gordon H, De Luca V (2008) The leaf epidermome of *Catharanthus roseus* reveals its biochemical specialization. Plant Cell 20(3):524–542. doi: 10.1105/tpc.107.056630
73. Goddijn OJ, de Kam RJ, Zanetti A, Schilperoort RA, Hoge JH (1992) Auxin rapidly down-regulates transcription of the tryptophan decarboxylase gene from *Catharanthus roseus*. Plant Mol Biol 18(6):1113–1120
74. Quint M, Gray WM (2006) Auxin signaling. Curr Opin Plant Biol 9(5):448–453. doi: 10.1016/j.pbi.2006.07.006
75. Benjamins R, Scheres B (2008) Auxin: the looping star in plant development. Annu Rev Plant Biol 59:443–465. doi:10.1146/annurev.arplant.58.032806.103805 (Annual Review of Plant Biology. Annual Reviews, Palo Alto)
76. Lee-Parsons CWT, Erturk S (2005) Ajmalicine production in methyl jasmonate-induced *Catharanthus roseus* cell cultures depends on Ca^{2+} level. Plant Cell Rep 24(11):677–682. doi:10.1007/s00299-005-0026-0
77. Oksman-Caldentey KM, Inze D (2004) Plant cell factories in the post-genomic era: new ways to produce designer secondary metabolites. Trends Plant Sci 9(9):433–440. doi: 10.1016/j.tplants.2004.07.006

78. Dafny-Yelin M, Tzfira T (2007) Delivery of multiple transgenes to plant cells. Plant Physiol. doi:10.1104/pp.107.106104
79. Lin L, Liu Y-G, Xu X, Li B (2003) Efficient linking and transfer of multiple genes by a multigene assembly and transformation vector system. Proc Natl Acad Sci U S A. doi: 10.1073/pnas.0931425100
80. Hadi MZ, McMullen MD, Finer JJ (1996) Transformation of 12 different plasmids into soybean via particle bombardment. Plant Cell Rep 15(7):500–505
81. Clarkson C, Staerk D, Hansen SH, Smith PJ, Jaroszewski JW (2006) Identification of major and minor constituents of *Harpagophytum procumbens* (devil's claw) using HPLC-SPE-NMR and HPLC-ESIMS/APCIMS. J Nat Prod 69(9):1280–1288. doi:10.1021/np0601612

Adv Biochem Eng Biotechnol (2013) 134: 55–90
DOI: 10.1007/10_2013_183

Published Online: 7 March 2013

Effective Elicitors and Process Strategies for Enhancement of Secondary Metabolite Production in Hairy Root Cultures

Jian Wen Wang and Jian Yong Wu

Abstract This chapter reviews the various biotic and abiotic elicitors applied to hairy root cultures and their stimulating effects on the accumulation of secondary metabolites. Elicitors generally refer to the agents that stimulate the defense responses of plants. As a major response of plants to biotic and abiotic stress, the accumulation of secondary metabolites in plant tissue cultures can be stimulated by the elicitors. Among the many elicitors applied to hairy root cultures as well as plant cell suspension cultures, the most common and effective elicitors are fungal cell extracts, polysaccharides from fungal and plant cells, and heavy metal salts. With the crude fungal cell extracts, it is essential to observe the preparation conditions carefully for achieving reproducible effects. In addition to the chemical agents, UV-radiation, hyperosmotic stress and temperature shift have been shown effective for some plant species/metabolites. Elicitor type, dose, and treatment schedule are major factors determining the effects on the secondary metabolite production. In addition to the accumulation of products in roots, elicitor treatments often stimulate the release of intracellular products. Although elicitation is mainly effective to increase specific product yield on per unit mass of roots, the incorporation of nutrient feeding strategies can be applied to enhance the volumetric product yield. The integration of in situ product recovery from the roots/liquid medium is another synergistic strategy with the elicitor treatment to improve the process.

Keywords Abiotic stress · Elicitors · Fungal extract · Integrated process · Nutrient feeding · Polysaccharides

---

J. W. Wang
College of Pharmaceutical Sciences, Soochow University,
Suzhou 215123, China
e-mail: jwwang@suda.edu.cn

J. Y. Wu (✉)
Department of Applied Biology & Chemical Technology,
The Hong Kong Polytechnic University,
Hung Hom Kowloon, Hong Kong
e-mail: bcjywu@polyu.edu.hk; jian-yong.wu@polyu.edu.hk

Contents

1 Introduction

The development of plant tissue (including organ and cell) cultures for the production of secondary metabolites has been underway for more than three decades. Although the basic techniques for plant tissue cultures have become well established, their application to large-scale production is still limited to a few processes [77, 91, 120]. Various stimulation and process strategies have been exercised to improve secondary metabolite production in plant tissue cultures. Elicitation, the treatment of plant cells with biotic and abiotic elicitors, has been one of the most effective means for the enhancement of secondary metabolite production in hairy root cultures as well as cell suspension cultures [10, 33]. This strategy works on the basis that the accumulation of most secondary metabolites in plants is part of the defense responses of plants to pathogen infection and environmental stresses. The agents that induce plant defense responses are generally referred to as elicitors. This chapter provides an overview of various elicitors and the common methods of elicitor treatment for enhancement of secondary metabolite production in hairy root cultures, and the integration of elicitation with other process strategies such as nutrient feeding and in situ product recovery to enhance hairy root production.

2 Elicitors and Their Effects on Hairy Root Cultures

In plant science, elicitors usually refer to the extracellular signal compounds of plant cells that trigger or initiate plant defense responses and phytoalexin synthesis [21]. This definition signifies that the elicitor is a signal rather than a physiological effector. It excludes the abiotic factors that have also been used to enhance secondary metabolite production in plant cell cultures. The elicitors by this definition can be classified as endogenous and exogenous. Endogenous elicitors are originated from the plant host as a result of its interaction with the aggressor, which play important roles in the intracellular signal transduction system. Among the best characterized are the pectic oligosaccharides released from plant cell walls [oligogalacturonides (OGAs) from citrus pectin] and the intracellular signal compounds (or second messengers), such as systemin, salicylic acid (SA), and jasmonic acid (JA), and relatives such as methyl jasmonate (MeJA). Endogenous elicitors may also be involved in mediating plant responses to abiotic stresses. Exogenous elicitors originate from the microbial pathogen or the aggressor, mostly fungal and bacteria, and a few from viruses and insects. The best-known elicitor compounds include peptides, polysaccharides, and glycoproteins derived from microbial and plant cells. The elicitors for stimulating secondary metabolite production in plant tissue cultures are generally classified as biotic and abiotic based on their origin [110]. Biotic elicitors originate from microbial and plant sources; abiotic elicitors include various chemical and physical stressors such as UV-radiation, heavy metals, and osmotic shock. Table 1 provides a comprehensive list of various biotic and abiotic elicitors applied to hairy root cultures.

2.1 Biotic Elicitors

2.1.1 Live Bacterial Cells

Hairy root culture provides a convenient experimental system for studying plant–microbe interactions. The hairy root–rhizobia co-culture between a climbing legume *Pachyrhizus erosus* and *Rhizobium* spp. was established for the study of hairy root nodulation [8]. Plant–bacteria interaction can also be utilized to improve the production of secondary metabolites in hairy root cultures. A plant-growth–promoting rhizobacterium *Bacillus cereus* has been applied to stimulate the secondary metabolite accumulation of *Salvia miltiorrhiza* hairy roots [94]. *S. miltiorrhiza* is an important medicinal plant in Chinese medicine and (its root part) is an effective herbal drug for cardiovascular diseases and inflammation, of which diterpenoid tanshinones are major bioactive constituents. With 0.2 % (v/v) bacterial inoculum applied to the hairy root culture on day 0 and co-cultured for 28 days, the total tanshinone content of roots was increased by more than 12-fold (from 0.20 to 2.67 mg/g DW), and the volumetric tanshinone yield increased by more than sixfold (from 1.40 to 10.4 mg/L; Table 2). The biomass growth of the root was suppressed by early inoculation of the bacterial cells (day 0) but was enhanced slightly by late inoculation (day 21). The study demonstrated an effective hairy root–bacterium co-culture for improving the production of secondary metabolites in hairy root cultures. The enhanced tanshinone production in the hairy root–bacteria co-culture system was most likely attributed to the elicitor action of

Table 1 Biotic and abiotic elicitors and their target secondary metabolites in hairy root cultures

Elicitor type	Elicitor agents	Products	Plant species	References
Biotic elicitors				
Live bacteria	*Rhizobium* spp.	Root nodulation	*Pachyrhizus erosus*	[8]
	Bacillus cereus	Tanshinone	*Salvia miltiorrhiza*	[94]
Yeast elicitor	Yeast extract	Ginseng saponin	*Panax ginseng*	[41]
		Tanshinones and phenolic acids	*Salvia miltiorrhiza*	[12, 29, 36, 72, 95, 100]
		Artemisinin	*Artemisia annua*	[65]
		Tropane alkaloids	*Brugmansia candida*	[64]
		Total flavonoids	*Glycyrrhiza uralensis*	[107]
		Glucotropaeolin	*Tropaeolum majus*	[92]
Fungal elicitor	Crude extracts of *Fusarium conglutinans* mycelia	Thiophenes	*Tayetes patula*	[58]
	Crude extracts of *Phytophthora cinnamon* mycelia	Rosmarinic acid	*Ocimum basilicum*	[2]
	Mycelia homogenates of *Penicillium chysogenum*	Artemisinin	*Artemisia annua*	[52]
Polysaccharides (PS)	PS from *Bacillus cereus* cells	Tanshinone	*Salvia miltiorrhiza*	[113]
	PS from *Colletotrichum gloeosporioides* mycelia	Artemisinin	*Artemisia annua*	[87]
	Oligogalacturonides	Artemisinin	*Artemisia annua*	[104]
	Heptasaccharide and octasaccharide	Ginseng saponin	*Panax ginseng*	[118, 119]
	Chitosan	Artemisinin	*Artemisia annua*	[65]
		Daidzein Genistein	*Psoralea corylifolia*	[74]
		Indigo	*Polygonum tinctorium*	[9]
Lipids	Cerebroside C from endophytic *Fusarim* sp.	Artemisinin	*Artemisia annua*	[88]

(continued)

Table 1 (continued)

Elicitor type	Elicitor agents	Products	Plant species	References
Glycoproteins	Microalgal glycoproteins extracted from *Scenedesmus obliquus*	*Cis*-C_{13}-spiroketal enol ether epoxide	*Tanacetum parthenium*	[78]
	Polysaccharide–protein fraction of *Bacillus cereus*	Tanshinones	*Salvia miltiorrhiza*	[113]
Signal molecules				
ROS and NO	H_2O_2 Methyl viologen (O_2^-cgenerator)	Tanshinones	*Salvia miltiorrhiza*	[11]
	H_2O_2	Sesquiterpene solavetivone, lubimin	*Hyoscyamus muticus*	[56]
	NO donor: SNP	Artemisinin	*Artemisia annua*	[88, 114]
Jasmonic acid (JA) and salicylic acid (SA) relatives	JA	Azadirachtin	*Azadirachta indica*	[70]
		Indole alkaloid	*Catharanthus roseus*	[61]
	Methyl jasmonate (MeJA)	Paclitaxel	*Taxus* × *media* var. *Hicksii*	[26, 79]
		Plumbagin	*Plumbago indica*	[28]
		Tanshinones	*Salvia miltiorrhiza*	[30, 50]
		Asiaticoside	*Centella asciatica*	[44]
		Ginsenoside	*Panax ginseng*	[60]
		Noradrenalin	*Portulaca oleracea*	[63]
		Artemisinin	*Artemisia annua*	[65]
		Glucotropaeolin	*Tropaeolum majus*	[92]
		Monoterpenoid indole alkaloids	*Ophiorrhiza pumila*	[99]
		Tropane alkaloid	*Hyoscyamus niger*	[109]

(continued)

Table 1 (continued)

Elicitor type	Elicitor agents	Products	Plant species	References
	SA	Tropane alkaloid	*Catharanthus roseus*	[62]
		Sonchuside A	*Cichorium intybus*	[54]
		Daidzein, genistein	*Psoralea corylifolia*	[74]
		Lubimin, solavetivone	*Hyoscyamus muticus*	[56]
		Ginseng saponin	*Panax ginseng*	[41]
		Tropane alkaloids	*Brugmansia candida*	[64]
		Tropane alkaloids	*Atropa belladonna*	[49]
		Monoterpenoid indole alkaloids	*Atropa belladonna*	[99]
	Acetylsalicylic acid (ASA)	Glucotropaeolin	*Tropaeolum majus*	[92]
		Ginseng saponin	*Panax ginseng*	[41]
Abiotic elicitors				
Heavy metal ions	Ag^+ ($Ag_2S_2O_3$)	Tanshinone	*Salvia miltiorrhiza*	[29, 30, 105]
	Ag^+ ($AgNO_3$)	Silymarin	*Silybum marianum*	[42]
		Scopolamine Hyoscyamine	*Brugmansia candida*	[64]
	Cd^{2+} ($CdCl_2$)	Lubimin 3-Hydroxylubimin	*Datura stramonium*	[27]
		Scopolamine Hyoscyamine	*Brugmansia candida*	[64]
	$CuCl_2$	Plumbagin	*Plumbago indica*	[28]
	$CuSO_4$	Umbelliferone Scopoletin	*Pharbitis nil*	[101]
		Lubimin 3-Hydroxylubimin	*Datura stramonium*	[27]
	$VOSO_4$	Thiarubrine A	*Ambrosia artemisiifolia*	[3]
		Total ginsenoside	*Panax ginseng*	[60]
	Selenium (selenite) Nickel ($NiSO_4$)	Ginseng saponin	*Panax ginseng*	[40]

(continued)

Table 1 (continued)

Elicitor type	Elicitor agents	Products	Plant species	References
Light and UV radiation	Light of different spectral ranges (fluorescent, metal halide, blue, red and blue plus red light)	Ginseng saponin	*Panax ginseng*	[102]
	Light 385–790 nm	Artemisinin	*Artemisia annua*	[89]
	Continuous light (cool-white fluorescent lamps, 40 μM m^{-2} s^{-1}) and in the ccontinuous dark	Diterpenoids (ferruginol, salvipisone, aethiopinone, and 1-oxoaethiopinone) Two ursenetype triterpenoids	*Salvia sclarea*	[48]
	Light intensity 0–3,000 Lux	Artemisinin	*Artemisia annua*	[51]
	Light 280–315 nm at 9,000 μW/cm^2	Total terpenoid indole alkaloids, lochnericine, strictosidine	*Catharanthus roseus*	[4, 62]
Temperature shift	Low temperature (19.5 °C)	Linolenic acid Indole alkaloids	*Catharanthus roseus*	[83]
	High temperature (40 °C, 45 °C, 50 °C)	Pigment	*Beta vulgari*	[82]
	Temperature shift (13 °C/20 °C, 20 °C/13 °C and 25 °C/25 °C for 16/8 h)	Ginsenoside	*Panax ginseng*	[93]
Osmotic stress	NaCl	Ginseng saponin	*Panax ginseng*	[40]
		Hyoscyamine	*Datura stramonium*	[43]
	Mannitol, sodium chloride, potassium chloride, cadmium chloride, and PVP K-30	Ajmalcine	*Catharanthus roseus*	[81]
	Sorbitol	Tanshinones	*Salvia miltiorrhiza*	[72, 95]
		Monoterpenoid indole alkaloids	*Ophiorrhiza pumila*	[99]

Table 2 Root growth and tanshinone production of *S. miltiorrhiza* hairy root cultures with various bacteria treatments

Treatment	Root dry weight (g/L)	Tanshinone content (mg/g DW)	Tanshinone yield (mg/L)
Live bacteria (bacteria inoculation vol. % and time)			
Control (no bacteria)	7.07 ± 0.72	0.20 ± 0.02	1.40
0.02 %, day 0	4.25 ± 0.13	1.98 ± 0.12	8.41
0.2 %, day 0	3.90 ± 0.27	2.67 ± 0.16	10.4
0.2 %, day 14	6.82 ± 0.45	1.25 ± 0.05	8.54
0.2 %, day 21	7.10 ± 0.21	0.79 ± 0.07	5.59

28-day overall culture period
Note The tanshinone content is the total content of three tanshinone species, cryptotanshinone (CT), tanshinone-I (T-I), and tanshinone-IIA (T-IIA)

some compounds released by the bacteria. Our later study showed that a polysaccharide–protein complex secreted by the bacteria to the culture medium stimulated the tanshinone accumulation significantly and also promoted hairy root growth [113].

2.1.2 Fungal (and Yeast) Elicitors

Various types of fungi have been used as the sources of fungal elicitors including pathogenic and endophytic fungi as well as fungi unrelated to plants. The fungal elicitors applied to hairy root cultures were mostly crude extracts of mycelial biomass or culture filtrates. The fungal elicitor for enhancing thiophene production of *Tayetes patula* hairy roots was prepared from *Fusarium conglutinans* cultured in potato-dextrose medium for 15 days. The fungal mycelium filtered from the culture broth was homogenized, and autoclaved at 121 °C for 30 min. The homogenate was filtered and the filtrate was concentrated by evaporation at 70–80 °C to the final elicitor preparation, which was quantified by the total carbohydrate content [58]. The content of 5-(4-acetoxy-l-butenyl)-2,2′-bithiophene and 5-(buten-l-enyl)-2,2′-bithiophene was increased in the hairy roots elicited by the elicitor at 0.124-mg carbohydrate/ml medium after an incubation period of 30 and 48 h. An elicitor preparation extracted from *Phytophthora cinnamon* increased the production of rosmarinic acid from *Ocimum basilicum* hairy roots by 2.67-fold [2]. In *Artemisia annua* hairy root cultures, the autoclaved mycelial homogenate of *Penicillium chysogenum* increased the artemisinin production in hairy root cultures by 1.2-fold over that of the control [52].

A yeast elicitor (YE) was isolated from yeast extract by ethanol precipitation [12]. Briefly, 50 g of the yeast extract was dissolved in 250 ml of distilled water. Ethanol was added to the solution at 80 % (v/v) for precipitation for 4 days at 4 °C. The gummy precipitate attained was re-dissolved in 250-mL distilled water and subjected to another round of ethanol precipitation. The second precipitate was collected and re-dissolved in 200-mL distilled water as the final YE preparation to be applied to the *S. miltiorrhiza* hairy root cultures. In that study, YE stimulated the production of both phenolic acids and tanshinones, and also improved the growth of hairy roots. A similar YE preparation was also applied to *Panax ginseng* hairy root cultures, increasing the total ginseng saponin content of hairy roots approximately 1.17 times, but inhibited hairy root growth by 20 % on a dry weight basis [41].

The generally effective elicitation and relatively simple preparation of the crude fungal elicitors are favorable factors for their wide applications. However, their undefined and complex chemical composition can cause difficulties in understanding which constituents are responsible for the elicitor activity, and in achieving reproducible effects from batch to batch or comparable effects reported by different research groups. These problems may be overcome by careful and strict control of the preparation procedures and conditions.

2.1.3 Polysaccharides from Microbial and Plant Cells

Polysaccharides and oligosaccharides isolated from plant and fungal cell walls have been widely used as elicitors in hairy root cultures. An endophytic fungus *Colletotrichum gloeosporioides* of *A. annua* has been used as a source of polysaccharide elicitors [87]. The polysaccharide isolated from the mycelium extract was acid-hydrolyzed to lower molecular fraction and oligosaccharide monomers were separated from the hydrolysate by gel filtration chromatography on a Sephadex G25 column. An oligosaccharide (MW < 2500) was isolated and applied as an elicitor to the *A. annua* hairy root culture. When the hairy root culture was treated with the elicitor at 0.4 mg/mL on day 23 (in the later growth phase) and further cultured for 4 days, the artemisinin yield was enhanced to a maximum of 13.5 mg/L, a 51.6 % increase over the control. OGAs, the pectic fragments of plant cell walls, also induced the artemisinin biosynthesis in *A. annua* hairy root cultures [104]. An OGA fraction was isolated from pectinase-hydrolyzed polygalacturonic acid using Sephadex G-10 column chromatography and was applied to stimulate the artemisinin accumulation. When 16-day-old hairy root cultures were exposed to the OGA elicitor (60 μg/mL) for 4 days, the artemisinin production reached a maximum of 11.3 mg/L, 55.2 % over the control. In *P. ginseng* hairy root culture, two plant-derived oligosaccharides (heptasaccharide and octasaccharide) from *Paris polyphylla* var. *yunnanensis* increased the saponin content of roots more than threefold [119]. To obtain the oligosaccharides, the dry rhizome powder was extracted with hot water and subjected to ethanol precipitation (to remove large polysaccharides). The remaining extract was purified by a Bio-Gel P-2 chromatographic column, giving two elicitors, heptasaccharide and octasaccharide (Fig. 1) [118].

The tanshinone accumulation of *S. miltiorrhiza* hairy roots was enhanced about sevenfold (1.59 mg/g vs. 0.19 mg/g) by a crude polysaccharide (designated as BPS) isolated from the hot water extract of *B. cereus* cells by ethanol precipitation [113]. BPS was a polysaccharide–protein complex containing about 27 % protein,

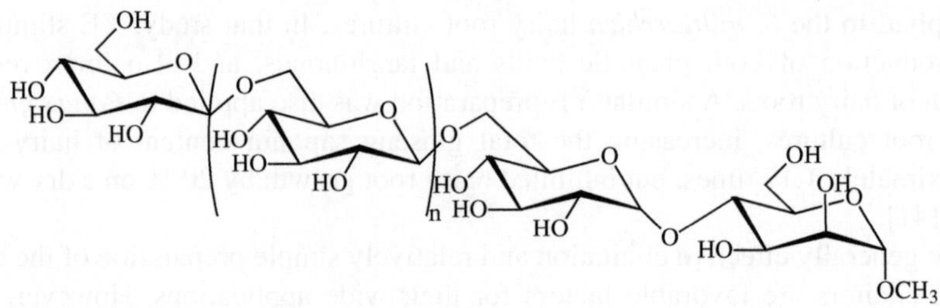

n=4: heptasaccharide
n=5: octasaccharide

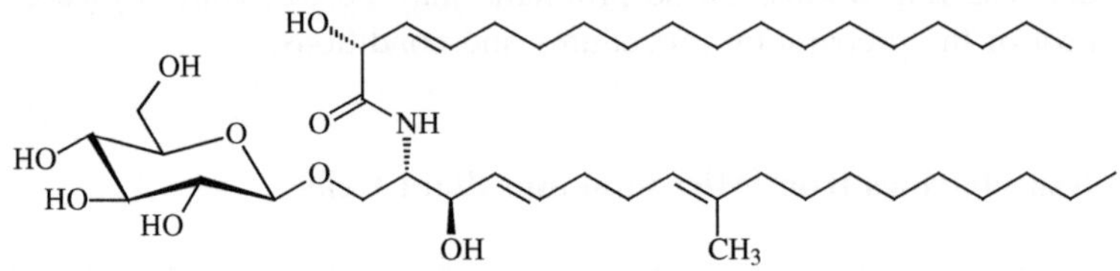

cerebroside C

Fig. 1 Chemical structures of some oligosaccharide and cerebroside elicitors

which was beneficial for root growth. Chitosan, an acetylated β-1,4-linked D-glucosamine polymer from fungal and plant cell walls, was also used as an elicitor in many previous studies. Chitosan (2 mg/L) was found to induce daidzein (2.78 % DW) and genistein (0.279 % DW) levels in hairy roots [74]. Putalun et al. [65] found that artemisinin production by *A. annua* hairy roots was increased sixfold to 1.8 μg/mg by adding 150 mg/L chitosan and with MeJA or YE increased by 1.5-fold to 0.9 μg/mg. The production of indigo in hairy root cultures of *Polygonum tinctorium* increased slightly with the addition of chitosan at 200 mg/L and pectinase at 20 U/L [9].

2.1.4 Lipid and Protein Elicitors

Some lipophilic compounds derived from bacteria, fungal pathogens, or endophytes also act as elicitors of plant defense response. Pathogen-derived glycosphingolipid cerebrosides have been reported as novel elicitors not only to induce rice phytoalexins and pathogenesis-related protein syntheses in rice suspension cultures but also as conferring disease resistance on rice blast fungus in paddy fields [45]. We have studied the enhanced accumulation of artemisinin in *A. annua* hairy roots induced by cerebroside C (Fig. 1) from endophytic *Fusarim* sp. [88]. The dry fungal mycelium was extracted successively with four volumes of 1:1 $CHCl_3$/MeOH by standing overnight at room temperature. The concentrated extract was soaked in 1.5 volumes of MeOH with reflux for 1 h and was then gradually cooled to −20 °C and allowed to stand for another 24 h to precipitate the lipids. The crude lipids were first separated by chromatography on a silica gel column, which was eluted sequentially with $CHCl_3$/MeOH mixtures of increasing polarity. Based on activity-guided fractionation by gel filtration on a Sephadex LH-20 column, a cerebroside elicitor was obtained, which stimulated artemisinin production by 70 %.

Glycoproteins have also been applied as elicitors to hairy root cultures. The microalgal glycoproteins (MGPS) extracted from *Scenedesmus obliquus* stimulated the accumulation (240 % of control) of *cis*-C_{13}-spiroketal enol ether epoxide concentration in hairy roots of *Tanacetum parthenium* and also enhanced the release of this metabolite to the culture medium [78]. After 30-day culture, alga cells and maternal cell walls were sedimented by centrifugation at 800 g for 10 min at 4 °C. Glycoproteins were precipitated from the supernatant with tannin solution and were recovered by freezing the precipitate with caffeine at −20 °C. MGPS were precipitated with acetone and the remaining solvent was removed under reduced pressure. MGPS stimulated the secondary metabolism of *T. parthenium* hairy roots as other common elicitors such as YE and $AgNO_3$, but had no inhibitory effect on root growth.

2.2 Abiotic Elicitors

2.2.1 Heavy Metal Ions

Many metals, especially the heavy metals, can stimulate the biosynthesis of secondary metabolites in plant tissue cultures. Silver ion Ag^+ in a concentrated silver thiosulfate ($Ag_2S_2O_3$) solution was prepared by mixing $AgNO_3$ and $Na_2S_2O_3$ at a 1:4 molar ratio. When Ag^+ (15, 30 μM) was supplied to *S. miltiorrhiza* hairy root cultures 2–3 days before the stationary growth phase (e.g., day 18) and followed by further incubation for 2 weeks, the total tanshinone (cryptotanshinone, tanshinone, and tanshinone IIA) content was increased twofold over the control but the root growth was suppressed to nearly 30 % lower dry weight [105]. When the *S. marianum* hairy root culture was fed with Ag^+ at 0.2–2 mM and harvested 72 h thereafter, the highest content of silymarin (0.56 mg/g DW) was obtained with 2 mM Ag^+ [42]. Ag^+ elicitation also induced the release of useful products into the culture medium. After 24 h, 1.0 mM $AgNO_3$ decreased the accumulation of scopolamine and hyoscyamine in hairy root cultures of *Brugmansia candida*, but significantly increased the release of scopolamine into the medium. After 48 h, $AgNO_3$ elicited overproduction of two tropane alkaloids in hairy roots [64]. In our previous study [30], Ag^+ was found to stimulate the activity of both the key enzyme 3-hydroxy-3-methylglutaryl CoA reductase (HMGR) in the mevalonate (MVA) pathway and 1-deoxy-D-xylulose 5-phosphate synthase (DXS) in the non-MVA, 1-deoxy-D-xylulose 5-phosphate (DXP) pathway, leading to tanshinone production.

This suggests that Ag^+ can trigger the defense responses and regulation on the cross-talk between the MVA and non-MVA isoprenoid pathways in the *S. miltiorrhiza* roots. The osmotic stress and the toxicity produced by the Ag^+ salt could probably explain the release of metabolites as a result of cell lysis. Moreover, the role of Ag^+ as an ethylene-blocking agent may also be counted [106], and ethylene could be involved in the negative effects on regulation of the pathway for some secondary metabolites.

Although cadmium (Cd) is a heavy metal that is toxic and detrimental to plant growth and development, Cd^{2+} has been used as a potent elicitor to stimulate the production of secondary metabolites in hairy root as well as plant cell cultures such as ajmalicine [116], podophyllotoxin [71], and tanshinones [112]. In hairy root culture of *Datura stramonium*, Cd^{2+} at 1 mM elicited the production of sesquiterpenoid lubimin and 3-hydroxylubimin, which were undetectable in unelicited cultures. The treatment with Cd^{2+} resulted in no detectable increase in the alkaloid content of the roots but in a notable increase in the release of alkaloids from roots to the medium after 40–60 h (up to 50–75 % of the total alkaloid) [27]. In *B. candida* hairy root cultures [64], $CdCl_2$ at 1.0–2.0 mM severely inhibited the growth of hairy roots but significantly stimulated the accumulation of two tropane alkaloids, hyoscyamine and scopolamine, and their release to the liquid medium (up to 24-fold). Although Cd is not a redox active metal to generate ROS directly, it often causes the plants to accumulate ROS and undergo oxidative stress [68]. The Cd-induced oxidative burst was shown to facilitate the release of secondary compounds into the culture medium [59, 121].

$CuCl_2$ elicitation caused a significant increase in intracellular plumbagin content (6.99 ± 0.33 mg g/L DW, 1.3-fold higher than control) at the dose of 300 mg/L on day 1 without significant alteration of biomass of *Plumbago indica* hairy roots [28]. The addition of $CuSO_4$ (64 μg/mL) to 21-day-old cultures of *Pharbitis nil* hairy roots resulted in the increased release of coumarin derivatives umbelliferone and scopoletin into medium. After the addition of $CuSO_4$, the umbelliferone content in the medium increased with time, reaching a peak in 16 h, but the content of skimmin, a *β*-D-glucopyranoside of umbelliferone in the hairy roots decreased to a minimal level after 8 h, and then increased to a level higher than the basal level after 24 h, suggesting that $CuSO_4$ triggered the endogenous biosynthetic pathway to supply skimming continuously [101].

Vanadyl sulphate ($VOSO_4$) has been found very effective for enhancement of paclitaxel, 10-deacetylbaccatin, and rosmarinic acid accumulation in different plant cell cultures [25, 31]. In hairy root cultures of *Ambrosia artemisiifolia*, an eightfold increase of thiarubrine A production (569 μg/g DW) was obtained when the 16-day-old culture was challenged with 50 mg/L $VOSO_4$ for 72 h [3]. When 16.3 mg/L $VOSO_4$ was added at the beginning (day 1) or on day 15 of *P. ginseng* hairy root culture, a significant reduction of the root growth was observed. However, when added on day 25, $VOSO_4$ only reduced the growth slightly and increased the total ginsenoside content by more than 1.6-fold [60].

Unlike the above-mentioned metal ions, selenium (Se) and nickel (Ni) are essential trace elements for plant growth, and have also been shown to enhance growth and saponin biosynthesis in *P. ginseng* hairy roots [40]. The addition of selenite (0.5–5 mM) at the time of inoculation caused strong growth inhibition on roots but no significant effect on the saponin biosynthesis. However, when 0.5 mM selenite was added after 21 days of culture, the root biomass was slightly higher than that observed in the control, and the saponin content and productivity were increased to about 1.31 and 1.33 times the control levels, respectively. The addition of 20 μM $NiSO_4$ resulted in a ginseng saponin content of 91.6 ± 9.8 mg/g (1.20 times the control value) and a productivity of 597 mg/L (1.23 times the control value), with no inhibitory effects on hairy root growth. Compared with other biotic elicitors and heavy metal salts, treatment with selenium and $NiSO_4$ as elicitors has the advantages of lower preparation cost, easier preparation, and no growth inhibition.

2.2.2 Light and Ultraviolet Radiation

Light has been shown to affect the formation of secondary metabolites including flavonoids, anthocyanins, gingerol, and zingiberene in callus and cell cultures [1, 47, 117]. The study of Yu et al. (2005) in *P. ginseng* hairy roots also showed that the exposure of hairy roots to different light spectral ranges affected growth and metabolite biosynthesis [102]. Although the root biomass was lower in the cultures grown under fluorescent irradiation (50 μmol m^{-2} s^{-1} per day) than dark incubation, the fluorescent irradiation enhanced the accumulation of ginsenosides Rg and Rb. Based on these results, a two-stage culture process, a dark stage for root growth followed by a fluorescent stage, is favorable for ginsenoside production. Another group reported the effects of light ranging from 385 to 790 nm on the growth and artemisinin biosynthesis of *A. annua* hairy roots [89]. The roots were incubated on a rotary shaker (120 rpm) at 25 ± 1 °C under 16 h light (cool-white, red, blue, yellow, and green fluorescence lamps) with the intensity of 25 μmol m^{-2} s^{-1} per day. After 30 days, red light at 660 nm was among the most favorable light source for promoting root biomass growth (5.73 g/L) and the artemisinin accumulation (31 mg/g DW), which were 17 and 67 %, respectively, higher than those obtained under white light. Light may be acting as a signal to activate the major enzymes in the biosynthetic steps for the secondary metabolites.

The quantitative and qualitative differences in diterpenoid and triterpenoid production in hairy roots of *S. sclarea* grown under continuous light (with cool-white fluorescent lamps at 40 μM m^{-2} s^{-1}) and in the dark were described by Kuźma et al. [48]. Light significantly increased diterpenoid concentration and slightly affected the triterpenoid profile in the hairy roots. The difference between triterpenoid compounds isolated from dark-grown and light-grown roots was attributed to the number of hydroxy groups in an ursine skeleton. In hairy root cultures of *A. annua*, growth rate of hairy roots increased significantly with the light intensity increased from 0 to 3,000 Lux. The biomass accumulation under continuous light (24 h) was higher than those under continuous darkness and other various periods (8, 12, 16 h) of light irradiation after 30 days. However, artemisinin content under continuous light was lower than that under the 16-h light followed by 8-h darkness [51], possibly due to the diversion of carbon flux from the secondary metabolic pathways under the continuous light.

UV light has been widely applied as an abiotic elicitor to stimulate the production of metabolites such as catharanthine vindoline and anthocyanin in plant cell cultures [32, 66]. UV-radiation not only stimulates but also induces the de novo synthesis of many secondary compounds in plants. In *C. roseus* hairy root cultures, roots were exposed to varying periods (5, 10, and 20 min) of UV-B light by placing the flasks on a UV illuminator with a wavelength range of 280–315 nm and a high-intensity setting of 9,000 $\mu W/cm^2$ at the surface. After exposure to the UV-B light, cultures were grown at 26 °C and 100 rpm in the dark for various periods from 0 to 168 h [4]. UV-B elicited an increase in the total terpenoid indole alkaloids (TIAs) concentrations in *C. roseus* hairy roots. A continuous increase in lochnericine levels for 48 h and a decrease in hörhammericine levels were observed simultaneously in response to 20 min of UV-B stress. Moreover, UV-B light exposure caused a large increase in transcripts of strictosidine synthase (STR) and tryptophan decarboxylase (TDC) in the hairy roots by 38 h. It has been reported that the UV stress response is related to the activation of the octadecanoid pathway in tomato leaves [14]. The inhibitors of the octadecanoid pathway including ibuprofen (IB), salicylhydroxamic acid (SHA), n-propyl gallate (PG), diethyldithiocarbamic acid (DIECA), SA were fed to *C. roseus* hairy roots 1 h before being exposed to UV-B stress. Only the higher concentration (1 mM) of DIECA and PG partially blocked the UV-induced TIA accumulation, indicating that UV-B could not activate the octadecanoid pathway in controlling TIA production in the roots [62].

2.2.3 Temperature Shift

Although temperature around 25 °C is normally used for hairy root cultures, lowering the cultivation temperature (19.5 °C) increased the proportion of linolenic acid and the total content of indole alkaloids in *C. roseus* hairy roots [83]. Despite the modifications in membrane lipid composition, temperature had no significant effect on alkaloid release to the medium. Moreover, as the root growth was much lower at the lower temperatures, the volumetric yield of indole alkaloids was lower than at normal culture temperature at 24 °C. The exposure to higher temperatures has been shown to affect the efflux of pigment from *Beta vulgaris* hairy roots [82]. Exposure of 20-day-old hairy root culture to 40, 45, and 50 °C resulted in the release of about 5.4, 31, and 47 % in 30 min and about 13.4, 40.2, and 47.5 % in 60 min, respectively, but the root viability was significantly lower at 50 °C. When *P. ginseng* hairy root cultures were maintained in 5-L flask bubble bioreactors at differential temperatures (e.g. 13 °C/20 °C, 20 °C/13 °C, and 25 °C/25 °C for 16/8 h day and night cycles), the biomass and ginsenoside production were optimal under 20 °C/13 °C day (12 h)/night (8 h) cycle [93]. The study suggests that temperature shift between day and night cycles is important for root growth and accumulation of secondary metabolite in the hairy roots.

2.2.4 Osmotic Stress

Hyperosmotic stress is another effective abiotic elicitor for stimulating production of secondary metabolites in hairy root cultures. NaCl and sorbitol are two common agents used to create the hyperosmotic stress in plant tissue cultures. In *P. ginseng* hairy root cultures, NaCl added to the culture medium increased the ginseng saponin content at all treated concentrations (0.01–0.3 %, w/v) and inhibited the growth only at 0.3 % (w/v). NaCl at a suitable concentration led to a 1.15-fold increase in the ginseng saponin content without inhibitory effect on the hairy root growth [40]. In another report, several osmotic salts including mannitol, sodium chloride, potassium chloride, cadmium chloride, and polyvinyl pyrrolidone (PVP) K-30 were tested on the growth and ajmalicine accumulation of *C. roseus* hairy root cultures in Gamborg's B5/2 medium [81]. The ajmalicine accumulation was increased to 2.53 mg/g DW (182 % vs. the control) with 0.2 % (w/v) PVP and to 4.09 mg/g DW (227 % increase) with 4 g/L KCl. The maximum concentrations of ajmalicine released to the medium were 5.4, 1.74, 2.192, and 2.02 mg/L, respectively, with mannitol (200 mM), cadmium chloride (10 μM), PVP (0.4 %, w/v), and NaCl (6 g/L) as compared to 1.32 mg/L in the control. However, the root viability and biomass concentration were decreased with the increase in the concentration of mannitol and salts. NaCl at 1–5 g/L did not show a significant effect on the biomass of *D. stramonium* hairy roots, but with NaCl treatment of 1, 2 g/L, hyoscyamine content in roots was three times higher than the control (3.54 and 4.53 mg/g DW, respectively) [43]. Peroxidase (POD) activity is frequently used as a biochemical marker of responses to salt stress. With NaCl treatment at 100 mM, the total activity of POD in tomato hairy roots did not change significantly but the POD activity in the medium decreased. NaCl induced secretion of a highly basic POD and inhibition of the secretion of some acidic isoenzymes [80]. In horseradish hairy roots, the addition of NaCl was a useful measure to increase production of POD in the industry [84]. In hairy root cultures of red beet (*B. vulgaris*), the cultures were exposed to different salt stresses such as NaCl, KCl, $MgCl_2$, $CaCl_2$, and EDTA at 5, 10, or 15 mM [69]. KCl and $CaCl_2$ at 5 mM increased POD activity slightly but promoted the secretion of POD significantly (nearly 43 and 50 %, respectively) and, at higher concentrations, suppressed enzyme productivity. Other salts tested did not cause any effect on POD production and secretion.

In *S. miltiorrhiza* hairy root cultures, sorbitol at 30–100 g/L was applied on day 21, 2–3 days before the stationary phase, increasing both the root mass and the total tanshinone content of roots (Table 3). Sorbitol at 70 g/L led to the highest tanshinone content of 723.6 μg/g DW, about 4.5-fold of the control (162.0 μg/g DW), but the highest volumetric yield of total tanshinone was attained at 50 g/L sorbitol, 8.18 mg/L [4.8-fold of the control (1.69 mg/L)] [72].

Table 3 Effect of osmotic stress (sorbitol) and YE on tanshinone production

Treatment	Root weight (g dw/L)	Tanshinone content (μg/g DW)				TT Yield (mg/L)
		CT	T-I	T-IIA	TT	
Control	12.1 ± 0.41	17.4	45.2	83.9	146.4 ± 17.6	1.77
Sorbitol	13.7 ± 0.32	313.1	96.1	262.1	671.3 ± 24.1	9.20
YE	10.4 ± 0.50	620.7	69.7	42.0	732.4 ± 90.3	7.62
Sorbitol + YE	11.0 ± 0.27	1344.6	50.0	87.0	1481.6 ± 36.5	16.3

Sorbitol at 50 g/L and YE at 100 mg/L; 9-day overall culture period
YE yeast elicitor (crude polysaccharide fraction of yeast extract); the total tanshinone (TT) production is the total production of three tanshinone species, cryptotanshinone (CT), tanshinone-I (T-I), and tanshinone-IIA (T-IIA)

2.3 Signal Molecules of Plant Defense Responses

The elicitation of plant defense responses involves a cascade of events between the elicitor signal perception by a receptor at the plant cell surface and the activation of defense-related genes in the nucleus. Several early events occurring after elicitor perception are important for the elicitor signal transduction, such as the induction of membrane Ca^{2+} influx, and transient production of ROS and nitric oxide (NO) and biosynthesis of JA and SA. These molecules play signaling roles for activating the downstream defense responses such as lipid peroxidation, activation of defense-related genes, and synthesis of secondary metabolites [110].

2.3.1 ROS and NO

Transient production of ROS such as hydrogen peroxide (H_2O_2) and superoxide anion (O_2^-), known as the oxidative burst, is an early event of plant response to pathogen (fungal, bacterial, and viral) and elicitors [53, 55]. The event has also occurred in hairy root cultures in response to various biotic and abiotic elicitors [33]. Although the inoculation of 0.2 % (v/v) live *B. cereus* bacteria into *S. miltiorrhiza* hairy root cultures stimulated tanshinone production, it also induced a sharp and rapid increase in the H_2O_2 level, reaching a peak of about 6.8–100 times over the initial and control level within 1–1.5 h [94]. An oligosaccharide elicitor from *F. oxysporum* mycelium induced rapid production of H_2O_2 in *A. annua* hairy root cultures, reaching a sharp peak of 7.3 μM around 1 h, and another lower and broad peak (3.1 μM) around 6 h (Fig. 2) [115].

Exogenously supplied H_2O_2 and methyl viologen (MV, paraquat), a generator of O_2^-, were used to induce phytoalexin synthesis in plant cells [5, 38]. In *S. miltiorrhiza* hairy root cultures, MV at 20–150 μM stimulated the production of cryptotanshinone in a dose-dependent manner, whereas H_2O_2 (10 μM–2 mM) alone cannot trigger the biosynthesis of tanshinones [11]. The effects of H_2O_2 on sesquiterpene solavetivone and lubimin production were tested in *Hyoscyamus*

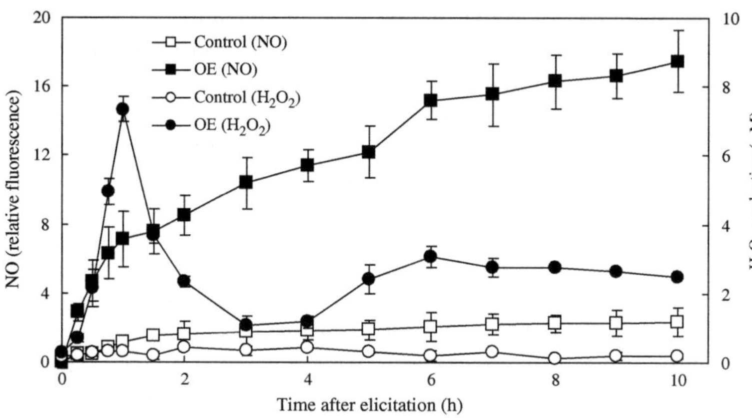

Fig. 2 Time course of oligosaccharide elicitor (OE)-induced nitric oxide (NO) and H_2O_2 accumulation in *Artemisia annua* hairy root cultures (OE at 0.3 mg total sugar/ml added to 20-day-old cultures. *Error bars* for standard deviations, $n = 3$)

muticus hairy root cultures [56]. H_2O_2 at 0.194–1.94 mM did not show any positive effect on sesquiterpene synthesis. Similar effects were observed in tobacco leaf panels in response to a challenge by incompatible bacteria *Pseudomonas corrugata* by Devlin and Gustine [20] that H_2O_2 was produced but was not necessary for activation of phytoalexin production. Moreover, H_2O_2 treatment alone could mimic the oxidative burst occurring in fungal elicitor-treated *C. roseus* cell cultures to stimulate alkaloid production [111]. In contrast, H_2O_2 (10 mM) could induce the accumulation of saponin in cultured ginseng cells [98]. Probably ROS play different roles in mediating the elicitation of defense responses and accumulation of secondary metabolites between cells and differentiated tissues.

NO as a signaling molecule in plant stress responses is involved in elicitor-induced production of secondary metabolites in plant cell and hairy root cultures [103]. The artificial NO donor sodium nitroprusside (SNP) was usually used as the exogenous source of NO, which promoted production of secondary metabolites such as catharanthine, baicalin, and hypericin production in plant cell cultures [96, 97, 108]. In *A. annua* hairy root cultures, SNP (10–100 μM) had promoting effects on root growth but no effect on artemisinin contents of the hairy roots, whereas the combination of SNP (10 μM) and an oligosaccharide elicitor stimulated higher artemisinin content in the hairy roots [114]. The stimulation of artemisinin accumulation was also observed when cerebroside C at 30 μg/mL was applied with the SNP treatment (10 μM) to *A. annua* hairy root cultures, suggesting that SNP could prime the hairy roots to be more sensitive and responsive to elicitation [88].

2.3.2 JA and SA

JA and its methylated derivative MeJA have been recognized as another class of signal transducer in elicitation of plant defense responses [18]. JA is derived from the octadecanoid pathway, which involves the peroxidation of linolenic acid by lipoxygenase (LOX) and the conversion of the lipid peroxide by allene oxide synthase (AOS). The accumulation of JA relatives occurs rapidly and transiently in plant tissues and cells after wounding [19] and exposure to fungal elicitors [35]. The relationship between elicitors and JA signaling pathways has been explored in hairy root cultures. Ag^+ at 2 mM stimulated the production of silymarin and related flavonolignan compounds in *Silybum marianum* hairy roots [42]. The LOX activity in JA signaling pathways was stimulated 433 % higher than the untreated hairy roots after 72-h elicitation and linoleic acid content was also enhanced. Both LOX activity and linoleic acid content in *S. marianum* hairy root cultures were stimulated after 72 h of YE (50 mg/L) treatment [36]. The fungal elicitor (1–10 mL/L) from *Rhizoctonia bataticola*, MeJA (10–200 μM) and *β*-cyclodextrin (10–200 mg/L) promoted the induction of LOX activity in *Solanum tuberosum* hairy root cultures [46]. The content of JA precursors 9-hydroxyoctadecadienoic acid and 9-hydroxyoctadecatrienoic acid was enhanced about 27-fold with the synergistic effect of the fungal elicitor and *β*-cyclodextrin.

Treatment of *Azadirachta indica* hairy root cultures with JA at 100 mM resulted in a six to ninefold increase in the production of azadirachtin, a limonoid antifeedant against a wide spectrum of insects [70]. A high content of asiaticoside (7.12 mg/g DW) was attained in *Centella asiatica* hairy roots in 3 weeks after application of 0.1 mM MeJA [44]. In putrescine *N*-methyltransferase overexpressing hairy root cultures of *H. niger*, MeJA treatment enhanced both polyamine and TIA biosynthesis [109]. Addition of JA also increased the specific yields of ajmalicine (80 %), serpentine (60 %), lochnericine (150 %), and hörhammericine (500 %) in *C. roseus* hairy roots [67]. The elicitation with JA (50 mg/L) on 20-day-old hairy root culture resulted in a 2.5-fold increase in the specific yield of the five alkaloids [57]. It has been reported that JA feeding to *C. roseus* hairy root cultures increased mRNA transcripts to activate several indole alkaloid biosynthesis genes for pathway induction [61]. Moreover, MeJA at 0.1 μM could trigger de novo accumulation of asiaticoside in *C. asiatica* hairy root cultures, an anti-inflammatory triterpene saponin that was not accumulated in the native hairy roots without the elicitation [44].

SA, an important signal molecule in systemic acquired resistance, has been used to enhance the production of secondary metabolites in hairy root cultures. SA caused a transient increase in the accumulation of a sesquiterpene lactone sonchuside A in *Cichorium intybus* hairy roots [54]. SA at 1 mM stimulated the accumulation of daidzein (2.2 % DW) and genistein (0.228 % DW) in *Psoralea corylifolia* hairy root cultures after 2-day elicitation, but higher concentrations of SA significantly reduced levels of daidzein and genistein in the roots [74]. The scopolamine content of *B. candida* hairy roots was enhanced with 1.0 mM SA, and hyoscyamine content increased with 0.01 and 1.0 mM SA. SA significantly increased the release of both alkaloids (2- to 12-fold) from roots to medium, but was detrimental to root growth, particularly after 72 h of treatment [64]. In hairy root cultures of *C. intybus,* SA at 100 μM did not influence biomass accumulation but caused a transient increase in accumulation of sonchuside A up to twofold compared with the control roots [54]. In *H. muticus* suspension cultures with 40 μM SA, lubimin production was increased by 50 %, and in hairy root cultures of the same species, lubimin and solavetivone production were increased by 5 and 48 %, respectively, with the addition of 4 μM SA [56]. The biomass growth of *P. ginseng* hairy roots was inhibited by SA at 0.1–1.0 mM, but the total ginseng saponin content was increased slightly at lower doses of 0.1–0.5 mM. Acetylsalicylic acid (ASA), a functional analogue of SA, appeared to be a potent elicitor for enhancing total ginseng saponin content in the roots [41].

SA and MeJA were tested on the production of noradrenalin in *Portulaca oleracea* hairy roots [63]. After 48-h treatment, the noradrenalin production was increased about eightfold with MeJA (200 μM) but not with SA (125–500 μM). In cotton hairy root cultures, SA did not increase the production of gossypol or its two methylated forms [22]. Although SA is not a universal inducer of metabolite production in hairy roots, it induced gene expression related to SA bioconversion and the biosynthesis of some classes of secondary metabolites. When 0.2 mM SA was added to *Atropa belladonna* hairy root culture, the tropane alkaloid production was not increased, but endogenous SA concentration decreased and was converted to methyl salicylate (MSA) through the reaction catalyzed by *S*-adenosyl-L-methionine: salicylic acid carboxyl methyltransferase (SAMT) [23]. During the elicitation, the expression of *A. belladonna* SAMT mRNA was induced by SA. In contrast, when hairy roots were treated with 2 mM SA, SA derivatives were not observed in the medium, and the tropane alkaloid release from the roots into the medium was increased up to 35 % of the total alkaloids after 24 h [49]. These results revealed that hairy root of *A. belladonna* exhibited distinct responses to SA depending on its concentration. The cDNA encoding UDP-glucose: baicalein 7-O-glucosyltransferase (UBGT), a key enzyme for transferring glucosyl moiety from UDP-glucose to the 7-hydroxy group of a flavone in hairy roots of *Scutellaria baicalensis* was cloned [37]. There was a rapid increase in the steady-state accumulation of UBGT mRNA after the wounding or treatment with SA, suggesting the involvement of glycosylation in regulation of the flavonoid synthesis in the hairy roots.

3 Elicitor Effects on Profile of Secondary Metabolites in Hairy Roots

In many cases, elicitors had differential effects on different secondary metabolites. For instance, YE as well as YE combined with sorbitol (osmotic stress) induced more significant increase in cryptotanshinone (CT) than the other two tanshinones (tanshinone I and IIA) in the *S. miltiorrhiza* hairy roots (Table 3) [72]. Similarly Ag^+ at 45 μM enhanced CT and tanshinone IIA yields but depressed tanshinone I yields in *S. miltiorrhiza* hairy root cultures [105]. Pectinase added to the late exponential phase of *C. roseus* hairy root cultures resulted in 150 % increase in tabersonine content but no significant changes in ajmalicine and serpentine content after 48 h [67]. The relative distribution of individual ginsenosides (four in the Rb group: Rb1, Rb2, Rc, and Rd, and three in the Rg group: Re, Rf, and Rg1) in *P. ginseng* hairy roots was also greatly influenced by the elicitation conditions [60]. In the presence of vanadyl sulphate (16.3 mg/L) at all treatment times (days 1, 15, and 25), the Rc levels increased and the total Rg percentage decreased, leading to a higher Rb/Rg ratio than in the control. The changes in the profile of active ingredients of the hairy roots may have an impact on the medicinal property of the roots.

Differential stimulation of secondary metabolite relatives in the same pathway by biotic and abiotic elicitors has also been observed in numerous other plant hairy root cultures, such as sesquiterpenes in *H. muticus* by MeJA and a fungal elicitor [75], indole alkaloids in *C. roseus* by pectinase and JA [67], and tropane alkaloids in *B. candida* by heavy metal ions and YE [64]. The effect of elicitors including MeJA, SA, YE, sodium chloride, and sucrose plus sorbitol on biosynthesis of TIA in hairy roots of *Ophiorrhiza pumila* was examined for the expression of strictosidine synthase (OpSTR), tryptophan decarboxylase (OpTDC), and NADPH: cytochrome P450 reductase (OpCPR) [99]. SA and YE treatments resulted in a repressive effect on the expression of both OpSTR and OpTDC, but MeJA exhibited no effect on their expression. In *S. miltiorrhiza* hairy root cultures, expression of two key genes for tanshinone production HMGR and DXR was up-regulated by MeJA and SNP [50]. In the tanshinone profile, the CT accumulation was most affected by MeJA elicitation at 100 μM, whereas CT and tanshinone IIA accumulation were more affected by SNP elicitation at μM than that of the other two componds. MeJA triggered the ROS burst and mediated to the increase of tanshinone production and up-regulation of related gene expression. However, SNP-induced increase of tanshinone production and expression of HMGR and DXR gene was probably not mediated by ROS. The stimulation effects on different secondary metabolite relatives by elicitors may be explained by the different physiological responses and different gene expression in the same pathway. However, a plausible explanation for this phenomenon is still not available owing to the complexity of the secondary metabolite biosynthetic pathways.

4 Effective Elicitation Methods

4.1 Optimal Dosage and Culture Age for Elicitation

The effect of an elicitor on secondary metabolite yield usually depends on the dosage and timing (or the age of culture) of treatment. For biotic elicitors, the effective and optimal doses are mostly in the range 50–300 μg/mL crude extracts, which are usually quantified by the total carbohydrate content. For purified compounds, signal compounds such as MeJA or SA, and abiotic elicitors of known compounds, the suitable doses are usually expressed in molar units, varying in a wide range from μM to mM levels (Table 4). In most published studies, the dose required for a culture system is determined and optimized based on dose–effect experiments. As most elicitor preparations are undefined crude extracts of biomass and the preparation protocols are not universal, the optimal dose derived by one group may be quite different from that by another, even when both use the same elicitor source for the same culture. Singh et al. [76] proposed the use of binding measurement to predict the elicitor dosage requirements for plant tissue cultures. The rationale is based on the assumption that elicitors bind to specific receptors on plant cells according to a saturation kinetics and an equilibrium between bound and free elicitors. According to this model, the amount or dose of elicitor required depends on the biomass concentration in the culture bioreactor. In addition, the elicitor dose should be more appropriately measured based on the amount of biomass (e.g., expressed by g elicitor/g biomass) but not on the culture volume. Such a dose unit has been used in a few previous studies in plant cell cultures, such as Funk et al. [24] for a YE and Bonfill et al. [6] for arachidonic acid, vanadyl sulfate, and MeJA. However, the elicitor doses applied in most previous studies have been chosen on an empirical basis or by trial and error, due partially to the crude and chemically undefined nature of most elicitor preparations.

The timing or culture age for elicitor treatment/feeding is another important factor of hairy root growth and metabolite production. Table 4 shows the elicitor doses and treatment time in numerous previous studies. Feeding in the middle or late growth phase is generally more favorable, whereas too early feeding may arrest the biomass growth and too late (during stationary phase) would not allow for sufficient contact time. For instance, Condori et al. [15] showed that production of resveratrol and the prenylated resveratrol analogues arachidin-1 and arachidin-3 in a medium of peanut hairy root cultures upon sodium acetate-mediated elicitation was highly dependent on the developmental stage. The highest yields were found during exponential growth. Similarly, the artemisinin production in *A. annua* hairy root cultures stimulated by a fungal elicitor was more significant in the exponential phase (on day 17) than in the stationary phase [86]. In contrast, the highest ginsenoside production in *P. ginseng* hairy root cultures was obtained when MeJA was applied in the stationary phase (day 28) [60]. It is possible that the plant cells are more prone to activate metabolic pathways associated with the production and secretion of these metabolites at certain developmental stages [73].

4.2 Multiple Elicitor Doses or Repeated Elicitor Treatment

A few studies have shown that multiple or repeated elicitor treatment over the culture period can enhance the secondary metabolite production more than a single treatment. In *S. miltiorrhiza* hairy root cultures, treatment with a YE twice, each at 100 mg/mL during the culture on days 18 and 20 enhanced the total diterpenoid tanshinone production (3.3 times volumetric yield) more significantly than treatment once on day 18 at 100 or 200 mM (2.3 times) [100]. The repeated treatment

Table 4 Effective elicitor dosage and treatment time (culture age) for enhancing secondary metabolite production in hairy root cultures

Products; plant species	Elicitors	Fold increase	Treatment time	Harvest time	Reference
Artemisinin; *Artemisia annua*	Mycelia extracts from *Colletotrichum* sp. (0.4 mg total sugar/mL)	1.44	Day 23	Day 27	[86]
	Oligogalacturonides (60 μg/mL)	1.55	Day 16	Day 20	[104]
	Oligosaccharide from *Fusarium oxysporum* (0.3 mg total sugar/mL)	1.66	Day 20	Day 24	[115]
	Cerebroside C (30 μg/mL)	1.7	Day 20	Day 24	[88]
Tanshinones; *Salvia miltiorrhiza*	Polysaccharide of *Bacillus cereus* (100 mg/L)	10.2	Day 18	Day 25	[113]
	Live *Bacillus cereus* cells 2.5 % (v/v)	12.3			
	Sterilized suspension from *Bacillus cereus* extract (1.0 g/L)	7.7	Day 14	Day 25	
	Yeast elicitor (100 mg/L)	3.8	Day 18	Day 30	[100]
	Ag^+ (30 μM)	1.2	Day 18	Day 20	[30]
	Bacillus cereus bacteria (0.2 %, v/v)	7.4	Day 0	Day 28	[94]
	Bacillus cereus bacteria extract (100 mg/L)	4.7			
	Bacillus cereus bacteria culture supernatant (21-day bacterial–root co-culture medium, 25 %, v/v)	4.6			
Asiaticoside; *Centella asiatica*	Methyl jasmonate (0.1 mM)	nd	Fourth week	Seventh week	[44]
Azadirachtin; *Azadirachta indica*	Dry cell powder of *Claviceps purpurea* (7 g/L)	4	Day 10	Day 25	[70]
	Jasmonic acid (100 mM)	3			
	Salicylic acid (100 mM)	4			

(continued)

Table 4 (continued)

Products; plant species	Elicitors	Fold increase	Treatment time	Harvest time	Reference
Ginseng saponins; *Panax ginseng*	Chitosan (250 mg/L)	nd	Day 25	Day 28	[60]
	Vanadyl sulfate (16.3 mg/L)				
	Methyl jasmonate (22.4 mg/L)				
	Selenium (0.5 mM)	1.33	Day 21	Day 24	[40]
	$NiSO_4$ (20 μM)	1.23			
	Sodium chloride (0.1 %, w/v)	1.15			
Plumbagin; *Plumbago indica*	Yeast carbohydrate fraction (1 mg/L)	1.2	Day 20	Day 23	[28]
	Chitosan (200 mg/L)	1.7	Day 20	Day 27	
	Manganese chloride (200 mg/L)	1.4	Day 20	Day 23	
	Copper chloride (300 mg/L)	1.3	Day 20	Day 21	
	Methyl jasmonate (80 μM)	2.0	Day 20	Day 23	
Total flavonoids; *Glycyrrhiza uralensis*	PEG8000 (2 %, w/v)	1.17	Day 20	Day 22	[107]
	Yeast extract (0.1 %, w/v)	1.6			
	PEG8000 (2 %, w/v) + Yeast extract (0.1 %, w/v)	2.3			
Daidzein and genistein; *Psoralea corylifolia*	Yeast extract (200 mg/L)	1.25 1.21	Day 18	Day 20	[74]
	Chitosan (10 mg/L)	1.8 1.47	Day 18	Day 23	
	Salicylic acid (1 mM)	1.42 1.2	Day 18	Day 20	
	Putrescine (50 mM)	1.94 1.2	Day 18	Day 23	
	Spermidine (25 mM)	1.5 1.78	Day 18	Day 23	

(continued)

Table 4 (continued)

Products; plant species	Elicitors	Fold increase	Treatment time	Harvest time	Reference
Silymarin; *Silybum marianum*	Ag^+ (2 mM)	3.7	Day 28	Day 32	[42]
	Yeast extract (50 μg/mL)	3.0	Day 30	Day 33	[36]
Gossypol and methylated gossypol; *Gossypium barbadense*	Methyl jasmonate (100 μM)	8 2–3	Day 10	Day 30	[22]
Spiroketal enol ether type diacetylenes	Yeast extract (0.08 %, w/v)	>3	nd	48–96 h after the treatment	[78]
Tanacetum parthenium	$AgNO_3$ (0.1 mM)				
	Microal glycoproteins from *Scenedesmus obliquus* (0.02 %, w/v)				
Glucotropaeolin; *Tropaeolum majus*	Salicylic acid (0.2 mM)	2.5	Day 9	Day 10	[92]
	Acetylsalicylic acid (0.2 mM)	3			
	Methyl jasmonate (50 μM)	1.6			
	β-aminobutyric acid (0.2 mM)	2			
	Yeast extract (0.05 %, w/v)	1.4			
	Acetylsalicylic acid (0.2 mM) + Methyl jasmonate (50 μM)	3.5	Day 9	Day 12	
	Acetylsalicylic acid (0.2 mM) + Yeast extract (0.05 %, w/v)				
Trans-resveratrol; *Arachis hypogaea*	Sodium acetate (10.2 mM)	30	Day 6	Day 7	[15]
		44	Day 9	Day 10	
		99	Day 12	Day 13	
		42	Day 15	Day 16	

nd Not detected

of the *S. miltiorrhiza* hairy root cultures with YE and sorbitol (hyperosmotic stress) in combination has also been shown more effective than a single treatment to stimulate the tanshinone production [95]. The larger increase in the volumetric product yield with multiple doses was partially attributed to less growth depression than with a single larger dose as well as a higher secondary metabolite accumulation in the roots. Similarly, repeated feeding of MeJA (100 μM) to *Taxus chinensis* cell cultures was more effective than a single dose (300 μM) to enhance the taxane production [90].

4.3 Synergism Between Two Elicitors: Potentiation of Elicitation

In *S. miltiorrhiza* hairy root cultures, the pre-treatment of hairy root cultures with Ag^+ (30 mM) for 0.5–2 days prior to the addition of a YE (100 μg/mL) significantly enhanced YE-induced tanshinone accumulation after 2 days of YE addition [29]. Ag^+ pre-treatment did not have any significant effect on HMGR activity but promoted the activity of DXS which is a key enzyme for the non-MVA pathway in tanshinone biosynthesis. Another study from our group has shown that the tanshinone yield was increased by simultaneous application of YE and herperosmotic stress (created with 50 g/L sorbitol) much more than each applied separately (Table 3) [72]. In addition to sorbitol, polyethylene glycol (PEG, molecular weight 8,000) has also been used to raise the medium osmolality. A study has shown that combination PEG + YE was more effective than the use of PEG (2 %, w/v) and YE (0.1 %, w/v) separately to stimulate the production of flavonoids in *Glycyrrhiza uralensis* hairy root cultures [107].

The secondary metabolite production in plants can be enhanced or potentiated by incorporation of the signal molecules in the elicitation of defense responses such as SA and JA relatives with the elicitor treatment. In *P. indica* hairy root cultures, 80 μM MeJA enhanced chitosan-induced total plumbagin content, reaching the highest level (11.96 ± 0.76 mg/g DW, ~2.3-fold higher than control) [28]. Treatment of *Tropaeolum majus* hairy roots with the combinations of 0.2 mM acetylsalicylic acid (ASA) and 50 μM MeJA or 0.05 % YE was more effective in increasing glucotropaeolin content than treatment with these elicitors alone [92]. This additive or synergistic effect suggested that induction of the biosynthesis of some metabolites by various stresses involves the interaction of the SA- or MeJA-mediated pathway with other elicitor signaling components.

In *S. miltiorrhiza* hairy root cultures, *β*-aminobutyric acid (BABA), a nonprotein amino acid, has been shown especially effective, more than MeJA, to potentiate the induced tanshinone production by YE. Pre-incubation of the cultures with BABA or MeJA for 24 h or longer before the elicitor treatment was essential for the potentiation to come into play whereas the addition of BABA or MeJA simultaneously with the elicitor had no effect [30]. The protective effect of BABA on plants against microbial pathogens has been recognized more recently, and the role of BABA and mode of action are less well understood [39]. BABA itself has no antimicrobial activity in general but can potentiate specific plant defense mechanisms in response to pathogen invasion. The studies on the defense resistance of *Arabidopsis* plants against virulent pathogens suggested that BABA mainly potentiated the defense mechanisms associated with the SAR signal pathway but not with the JA pathway [122]. It has been suggested that BABA can prime the plant cells to be more sensitive and responsive to elicitation, although the mechanism of priming is still not fully understood [16].

In hairy root cultures of *A. annua*, the combination application of NO donor SNP (at 50 μM) and an oligosaccharide elicitor (at 0.3 mg total sugar/mL) increased the artemisinin content from 1.2 to 2.2 mg/g DW, to a maximum artemisinin yield of 28.5 mg/L, about twofold over that attained with the oligosaccharide treatment alone. In the absence of oligosaccharide elicitor, SNP at 10, 50, and 100 μM enhanced the growth of hairy roots but had no effect on artemisinin synthesis [114]. In our recent work, we found that the combination of SNP (10 μM) and a fungus-derived cerebroside C (30 μg/mL) stimulated a higher artemisinin content in *A. annua* hairy root cultures [88].

4.4 Integration of Elicitation with Other Strategies

4.4.1 Elicitation and Precursor Feeding

Addition of metabolite precursors or their relatives is a useful strategy for increasing secondary metabolite production in plant cell and hairy root cultures. The combination of precursor feeding with elicitation may be a synergistic approach to increasing the yield of the target metabolite product, on the hypothesis that elicitation stimulates the biosynthetic activity of cells to metabolize the exogenously supplied precursors. In *T. majus* hairy root cultures, when two amino acid precursors for glucotropaeolin, phenylalanine, and cystein were added at 0.6 mM to 9-day old cultures, the glucotropaeolin content was increased twofold after 24-h treatment compared to the control [92]. Application of an SA analogue ASA for 24 h at 0.2 mM in combination with the precursors was even more effective to increase the glucotropaeolin content, leading to about a 4.8-fold increase in the glucotropaeolin yield, which was also higher than in the cultures treated with ASA alone. In hairy root cultures of *Taxus* × *media* var. *Hicksii*, the greatest paclitaxel yield (420 μg/g DW) was achieved after 3-week elicitation with 100 μM MeJA, and the 10-deacetybaccatin III content was not affected by the elicitor [26]. Supplementation with 100 μM of the precursor L-phenylalanine together with 100 μM of MeJA resulted in not only the enhancement of paclitaxel production from 40.3 (control) to 568.2 μg/L, but also induced the accumulation of 10-deacetybaccatin III (22.3 μg/L) after 2-week treatment. The combination of another precursor 100 μM p-amino benzoic acid with MeJA had a less pronounced effect on taxane production, leading to a paclitaxel yield of 221.8 μg/L [79]. Conversely, combined elicitation with 50 mg/L JA and precursor feeding with loganin at 40 mg/L or tryptamine at 20 mg/L did not further enhance the accumulation of indole alkaloids in *C. roseus* hairy root cultures, and conversely, addition of the precursors a day after elicitation significantly decreased the specific yields of tabersonine, ajmalicine, and serpentine [57]. More in-depth studies are needed to understand the biosynthetic pathway and to develop effective combined feeding schemes of precursors and elicitors for enhanced production of the target metabolites in hairy root cultures.

4.4.2 Elicitation and Nutrient Feeding

In principle, elicitation is to induce or stimulate the secondary metabolism of plant cells but is not favorable for cell growth (primary metabolism). Suppressed biomass growth has been observed in hairy root cultures treated by abiotic elicitors and biotic elicitors, especially at relatively larger doses [27, 64]. The adverse effect of elicitor treatment on the biomass growth can be alleviated by the feeding of growth nutrients including sucrose and other medium components. Hence, much higher productivity may be achieved by the integration of elicitation with a nutrient feeding scheme that can sustain biomass growth. Moreover, nutrient feeding allows for the application of multiple elicitor doses or repeated elicitor treatments over a longer culture period so as to enhance production even more dramatically. Our previous study has shown that the integration of Ag^+ elicitation and nutrient feeding (or medium renewal) is an effective strategy for improving tanshinone production in *S. miltiorrhiza* hairy root cultures [105]. Although addition of $Ag_2S_2O_3$ at 15–40 μM to the culture between 12 and 22 days post inoculation caused a dose-dependent inhibition of hairy root growth, sucrose feeding, or medium renewal before Ag^+ addition effectively prevented the growth inhibition and significantly enhanced the root biomass and volumetric tanshinone yields in culture. For sucrose feeding, a concentrated sucrose stock solution was added to the culture at a final concentration of 60 g/L (30 g/L for routine growth medium). For medium renewal, the spent medium in the culture flasks was replaced with fresh medium containing 60 g/L sucrose. The biomass yield was increased 76 % by sucrose feeding and more than 100 % by medium renewal. Although the nutrient feeding or medium renewal operation had a slight effect on the specific tanshinone content in roots, it increased the volumetric yield of total tanshinone significantly. Our results suggest the enhancement of tanshinone production by nutrient feeding was mainly due to the promotion of biomass growth.

More recently, an ultrahigh tanshinone yield (18.1 mg/g DW content in root and 145 mg/L volumetric yield) was achieved in *S. miltiorrhiza* hairy root cultures through repeated osmotic stress (OS) + YE stimulation and nutrient medium in an extended fed-batch culture process (i.e., 10 mL fresh medium with 50 g/L sorbitol and 25 mg/L YE, every 5 days from day 21 to day 60) [95]. Such a high tanshinone content of the hairy roots has never been reported before with *S. miltiorrhiza* roots from plant tissue culture or with field-grown plant roots (mostly $\leq$ 1.0 %) [85]. Elicitation (OS + YE treatment) for stimulating the biosynthesis to higher levels and nutrient feeding for sustaining the biomass growth are two complementary strategies for substantial enhancement of the secondary metabolite production.

4.4.3 Elicitation and in Situ Product Removal

The production of target metabolites can be limited by both feedback inhibition of accumulated synthesized metabolites and their degraded compounds. The introduction of an in situ product removal mechanism, such as a solid adsorbent or an extraction solvent, to the culture medium can often effectively induce product release from plant cells and increase productivity [13]. For in situ adsorption, the nonionic polymeric ion-exchange resins of Amberlite® XAD series (Rohm and Haas, Philadelphia, PA) have been used in previous studies in hairy root cultures, such as XAD-2 and XAD-4 for recovery of alkaloids in *Nacotiana glauca* hairy root cultures [34], XAD-7 for thiophene excretion in *Tagetes patula* hairy root cultures [7], and XAD-7, XAD-16 for integrated recovery of solavetivone in *H. muticus* hairy root cultures [17]. In *S. miltiorrhiza* hairy root cultures [100], the addition of X-5 macroporous polystyrene resin (2 g of X-5 resin in a nylon bag in each 200-mL flask) recovered a major portion of tanshinones from the roots, 70–94 %. Moreover, the combination of the X-5 adsorption with double YE elicitation (added to the culture on days 30 and 40 at 100 mg/L) led to a significant increase in the tanshinone yield. To prolong biomass growth and secondary metabolite production, we performed medium renewal every 10 days for three times (days 30, 40, and 50). With the combination of repeated medium renewal, YE elicitation, and resin replacement, the root biomass was increased to 30 g/L DW (vs. 8–10 g/L DW in batch mode) and the volumetric tanshinone yield to 87 mg/L (about a 15-fold increase), with 76 % adsorbed to the resin. The volumetric productivity of total tanshinone reached 1.46 mg/L day, more than 7 times that of the batch culture.

5 Concluding Remarks

Elicitation has been widely applied for enhancement of secondary metabolite production in hairy root cultures as well as in plant cell cultures. Among the many biotic and abiotic elicitors used in previous studies, the most common and effective elicitors for stimulating secondary metabolite production in hairy root cultures are crude extracts or partially purified polysaccharide fractions of fungal cells, JA (or MeJA), and heavy metal ions. The incorporation of nutrient feeding strategies with elicitation is often essential for achieving a high volumetric yield of the secondary metabolites in culture. Simultaneous or in situ product removal from the roots and/or liquid medium is another useful measure for improving production. These process strategies are more easily applied in hairy root cultures than cell suspension cultures because the roots are self-immobilized and retained within the culture vessel, allowing for liquid medium renewal (withdrawal and addition).

Acknowledgments We appreciate the financial support from the Hong Kong Government UGC (GRF Projects PolyU 5028/09P and PolyU 5036/10P), the Guangdong Provincial Government (Project 2011B090400272), and the National Natural Science Foundation of China (No. 81273487), PAPD and SRF for ROCS (No. K513201011).

References

1. Anasori P, Asghari G (2008) Effects of light and differentiation on gingerol and zingiberene production in callus culture of *Zingiber officinale* Rosc. Res Pharm Sci 3:59–63
2. Bais HP, Walker TS, Schweizer HP, Vivanco JA (2002) Root specific elicitation and antimicrobial activity of rosmarinic acid in hairy root cultures of *Ocimum basilicum*. Plant Physiol Biochem 40:983–995
3. Bhagwath SG, Hjortsø MA (2000) Statistical analysis of elicitation strategies for thiarubrine A production in hairy root cultures of *Ambrosia artemisiifolia*. J Biotechnol 80:159–167
4. Binder BYK, Peebles CAM, Shanks JV, San K-Y (2009) The effects of UV-B stress on the production of terpenoid indole alkaloids in *Catharanthus roseus* hairy roots. Biotechnol Prog 25:861–865
5. Bolwell GP, Davies DR, Gerrish C, Auh CK, Murphy TM (1998) Comparative biochemistry of the oxidative burst produced by rose and French bean cells reveals two distinct mechanism. Plant Physiol 116:1379–1385
6. Bonfill M, Palazón J, Cusidó RM, Joly S, Morales C, Piñol MT (2003) Influence of elicitors on taxane production and 3-hydroxy-3-methylglutaryl coenzyme A reductase activity in *Taxus media* cells. Plant Physiol Biochem 41:91–96
7. Buitelaar RM, Leenen EJTM, Geurtsen G, de Groot E, Tramper J (1993) Effects of the addition of XAD-7 and the elicitor treatment on growth, thiophene production, and excretion by hairy roots of *Tagetes patula*. Enzyme Microb Technol 15:670–676
8. Cadiz NM, Vivanco JM, Flores HE (2000) Coculture of *Pachyrhizus erosus* (L.) hairy roots with *Rhizobium* spp. In Vitro Cell Dev Biol-Plant 36:238–242
9. Chae YA, Yu HS, Song JS, Chun HK, Park SU (2000) Indigo production in hairy root cultures of *Polygonum tinctorium* Lour. Biotechnol Lett 22:1527–1530
10. Chandra S, Chandra R (2011) Engineering secondary metabolite production in hairy roots. Phytochem Rev 10:371–395
11. Chen H, Chen F (2000) Induction of phytoalexin formation in crown gall and hairy root culture of Salvia miltiorrhiza by methyl viologen. Biotechnol Lett 22:715–720
12. Chen H, Chen F, Chiu FCK, Lo CMY (2001) The effect of yeast elicitor on the growth and secondary metabolism of hairy root cultures of *Salvia miltiorrhiza*. Enzym Microb Technol 28:100–105
13. Choi JW, Cho GH, Byun SY (2001) Integrated bioprocessing for plant cell cultures. Adv Biochem Eng/Biotechnol 72:63–102
14. Conconi A, Smerdon MJ, Howe GA, Ryan CA (1996) The octadecanoid signalling pathway in plants mediates a response to ultraviolet radiation. Nature 383:826–829
15. Condori J, Sivakumar G, Hubstenberger J, Dolan MC, Sobolev VS, Medina-Bolivar F (2010) Induced biosynthesis of resveratrol and the prenylated stilbenoids arachidin-1 and

arachidin-3 in hairy root cultures of peanut: effects of culture medium and growth stage. Plant Physiol Biochem 48:310–318
16. Conratha U, Pieterse CMJ, Mauch-Mani B (2002) Priming in plant–pathogen interactions. Trends Plant Sci 7:210–216
17. Corry JP, Reed WL, Curtis WR (1993) Enhanced recovery of solavetivone from *Agrobacterium* transformed root cultures of *Hyoscyamus muticus* using integrated product extraction. Biotechnol Bioeng 42:503–508
18. Creelman RA, Mullet JE (1995) Jasmonic acid distribution and action in plants: regulation during development and response to biotic and abiotic stress. Proc Natl Acad Sci USA 92:4114–4119
19. Creelman RA, Tierney ML, Mullet JE (1992) Jasmonic acid/methyl jasmonate accumulate in wounded soybean hypocotyls and modulate wound gene expression. Proc Natl Acad Sci USA 89:4938–4941
20. Devlin WS, Gustine DL (1992) Involvement of the oxidative burst in phytoalexin accumulation and the hypersensitive reaction. Plant Physiol 100:1189–1195
21. Ebel J, Cosio EG (1994) Elicitors of plant defense responses. Int'l Rev Cytol 148:1–36
22. Frankfater CR, Dowd MK, Triplett BA (2009) Effect of elicitors on the production of gossypol and methylated gossypol in cotton hairy roots. Plant Cell, Tissue Organ Cult 98:341–349
23. Fukami H, Asakura T, Hirano H, Abe K, Shimomura K, Yamakawa T (2002) Salicylic acid carboxyl methyltransferase induced in hairy root cultures of *Atropa belladonna* after treatment with exogeneously added salicylic acid. Plant Cell Physiol 43:1054–1058
24. Funk C, Gügler K, Brodeliusa P (1987) Increased secondary product formation in plant cell suspension cultures after treatment with a yeast carbohydrate preparation (elicitor). Phytochemistry 26:401–405
25. Furmanowa M, Olędzka H, Sykłowska-Baranek K, Józefowicz KJ, Gieracka S (2000) Increased taxane accumulation in callus cultures of *Taxus cuspidate* and *Taxus* × *media* by some elicitors and precursors. Biotechnol Lett 22:1449–1552
26. Furmanowa M, Syklowska-Baranek K (2000) Hairy root cultures of *Taxus* × *media* var. *Hicksii* Rehd. as a new source of paclitaxel and 10-deacetylbaccatin III. Biotechnol Lett 22:683–686
27. Furze JM, Rhodes MJC, Parr AJ, Robins RJ, Withehead IM, Threlfall DR (1991) Abiotic factors elicit sesquiterpenoid phytoalexin production but not alkaloid production in transformed root cultures of *Datura stramonium*. Plant Cell Rep 10:111–114
28. Gangopadhyay M, Dewanjee S, Bhattacharya S (2011) Enhanced plumbagin production in elicited *Plumbago indica* hairy root cultures. J Biosci Bioeng 111:706–710
29. Ge XC, Wu JY (2005) Induction and potentiation of diterpenoid tanshinone accumulation in *Salvia miltiorrhiza* hairy roots by *β*-aminobutyric acid. Appl Microbiol Biotechnol 68:183–188
30. Ge XC, Wu JY (2005) Tanshinone production and isoprenoid pathways in *Salvia miltiorrhiza* hairy roots induced by Ag^+ and yeast elicitor. Plant Sci 168:487–491
31. Georgiev M, Kuzeva S, Pavlov A, Kovacheva E, Ilieva M (2006) Enhanced rosmarinic acid production by *Lavandula vera* MM cell suspension culture through elicitation with vanadyl sulfate. Z Naturforsch 61c: 241-244
32. Gläβgen WE, Rose A, Madlung J, Koch W, Gleitz J, Seitz HU (1998) Regulation of enzymes involved in anthocyanin biosynthesis in carrot cell cultures in response to treatment with ultraviolet light and fungal elicitors. Planta 204:490–498
33. Goel MK, Mehrotra S, Kukreja AK (2011) Elicitor-induced cellular and molecular events are responsible for productivity enhancement in hairy root cultures: an insight study. Appl Biochem Biotechnol 165:1342–1355
34. Green KD, Thomas NH, Callow JA (1992) Product enhancement and recovery from transformed root cultures of *Nicotiana glauca*. Biotechnol Bioeng 39:195–202
35. Gundlach H, Müller MJ, Kutchan TM, Zenk MH (1992) Jasmonic acid is a signal transducer in elicitor-induced plant cell cultures. Proc Natl Acad Sci USA 89:2389–2393

36. Hasanloo T, Sepehrifar R, Rahnama H, Shams MR (2009) Evaluation of the yeast-extract signaling pathway leading to silymarin biosynthesis in milk thistle hairy root culture. World J Microbiol Biotechnol 25:1901–1909
37. Hirotani M, Kuroda R, Suzuki H, Yoshikawa T (2000) Cloning and expression of UDP-glucose: flavonoid 7-O-glucosyltransferase from hairy root cultures of *Scutellaria baicalensis*. Planta 210:1006–1013
38. Jabs T, Tschöpe M, Colling C, Hahbrock K, Scheel D (1997) Elicitor-stimulated ion fluxes and O_2^- from the oxidative burst are essential components in triggering defense gene activation and phytoalexin synthesis in parsley. Proc Natl Acad Sci USA 94:4800–4805
39. Jakab G, Cottier V, Toquin V, Rigoli G, Zimmerli L, Métraux JP, Mauch-Mani B (2001) β-aminobutyric acid-induced resistance in plants. Eur J Plant Pathol 107:29–37
40. Jeong GT, Park DH (2006) Enhanced secondary metabolite biosynthesis by elicitation in transformed plant root system: effect of abiotic elicitors. Appl Biochem Biotechnol 129–132:436–446
41. Jeong GT, Park DH, Ryu HW, Hwang B, Woo JC, Kim D, Kim SW (2005) Production of antioxidant compounds by culture of *Panax ginseng* C.A. Meyer hairy roots: I. Enhanced production of secondary metabolite in hairy root cultures by elicitation. Appl Biochem Biotechnol 121–124:1147–1157
42. Khalili GM, Hasanloo T, Tabar SKK (2010) Ag^+ enhanced silymarin production in hairy root cultures of *Silybum marianum* L. Plant Omics 3:109–114
43. Khelifi L, Zarouri B, Amdoun R, Harfi B, Morsli A, Khelifi-Slaoui M (2011) Effects of elicitation and permeabilization on hyoscyamine content in *Datura stramonium* hairy roots. Adv Environ Biol 5:329–334
44. Kim OT, Bang KH, Shin YS, Lee MJ, Jung SJ, Hyun DY, Kim YC, Seong NS, Cha SW, Hwang B (2007) Enhanced production of asiaticoside from hairy root cultures of *Centella asiatica* (L.) Urban elicited by methyl jasmonate. Plant Cell Rep 26:1941–1949
45. Koga J, Yamauchi T, Shimura M, Ogawa N, Oshima K, Umemura K, Kikuchi M, Ogasawara N (1998) Cerebrosides A and C, sphingolipid elicitors of hypersensitive cell death and phytoalexin accumulation in rice plants. J Biol Chem 273:31985–31991
46. Komaraiah1 P, Reddy GV, Reddy PS, Raghavendra AS, Ramakrishna SV, Reddanna P (2003) Enhanced production of antimicrobial sesquiterpenes and lipoxygenase metabolites in elicitor-treated hairy root cultures of Solanum tuberosum. Biotechnol Lett 25: 593–597
47. Krewzaler F, Hahlbrock K (1973) Flavonoid glycosides from illuminated cell suspension cultures of *Petroselinum hortense*. Phytochemistry 12:1149–1152
48. Kuźma Ł, Skrzypek Z, Wysokińska H (2006) Diterpenoids and triterpenoids in hairy roots of *Salvia sclarea*. Plant Cell Tiss Organ Cult 84:171–179
49. Lee K-T, Hirano H, Yamakawa T, Kodama T, Igarashi Y, Shimomura K (2001) Responses of transformed root culture of *Atropa belladonna* to salicylicacid stress. J Biosci Bioeng 91:586–589
50. Liang Z-S, Yang D-F, Liang X, Zhang Y-J, Liu Y, Liu F-H (2012) Roles of reactive oxygen species in methyl jasmonate and nitric oxide-induced tanshinone production in *Salvia miltiorrhiza* hairy roots. Plant Cell Rep 31:873–883
51. Liu C, Guo C, Wang Y, Ouyang F (2002) Effect of light irradiation on hairy root growth and artemisinin biosynthesis of *Artemisia annua* L. Process Biochem 38:581–585
52. Liu C, Wang Y, Xu X, Ouyang F, Ye H, Li G (1999) Improvement of artemisinin accumulation in hairy root cultures of *Artemisia annua* L by fungal elicitor. Bioproc Biosyst Eng 20:161–164
53. Low PS, Merida JR (1996) The oxidative burst in plant defense: function and signal transduction. Physiol Plant 96:533–542
54. Malarz J, Stojakowska A, Kisiel W (2007) Effect of methyl jasmonate and salicylic acid on sesquiterpene lactone accumulation in hairy roots of *Cichorium intybus*. Acta Physiol Plant 29:127–132
55. Mehdy MC (1994) Active oxygen species in plant defense against pathogens. Plant Physiol 105:467–472

56. Mehmetoglu Ü, Curtise WR (1997) Effects of abiotic inducers on sesquiterpene synthesis in hairy root and cell suspension cultures of *Hyoscyamus muticus*. Appl Biochem Biotechnol 67:71–77
57. Morgan JA, Shanks JV (2000) Determination of metabolic rate-limitations by precursor feeding in *Catharanthus roseus* hairy root cultures. J Biotechnol 79:137–145
58. Mukundan U, Hjortso MA (1990) Effect of fungal elicitor on thiophene production in hairy root cultures of *Tayetes patula*. Appl Microbiol Biotechnol 33:145–147
59. Olmos E, Martinez-Solano JR, Piqueras A, Hellin E (2003) Early steps in the oxidative burst induced by cadmium in cultured tobacco cells (BY-2 line). J Exp Bot 54:291–301
60. Palazón J, Cusidó RM, Bonfill M, Mallol A, Moyano E, Morales C, Piñol MT (2003) Elicitation of different *Panax ginseng* transformed root phenotypes for an improved ginsenoside production. Plant Physiol Biochem 41:1019–1025
61. Peebles CAM, Hughes EH, Shanks JV, San KY (2009) Transcriptional response of the terpenoid indole alkaloid pathway to the overexpression of ORCA3 along with jasmonic acid elicitation of *Catharanthus roseus* hairy roots over time. Metab Eng 11:76–86
62. Peebles CAM, Shanks JV, San K-Y (2009) The role of the octadecanoid pathway in the production of terpenoid indole alkaloids in *Catharanthus roseus* hairy roots under normal and UV-B stress conditions. Biotechnol Bioeng 103:1248–1254
63. Pirian K, Piri K (2012) Effect of methyl jasmonate and salicylic acid on noradrenalin accumulation in hairy roots of *Portulaca oleracea* L. Int Res J Appl Basic Sci 3:213–218
64. Pitta-Alvarez SI, Spollansky TC, Giulietti AM (2000) The influence of different biotic and abiotic elicitors on the production and profile of tropane alkaloids in hairy root cultures of *Brugmansia candida*. Enzyme Microb Technol 26:252–258
65. Putalun W, Luealon W, De-Eknamkul W, Tanaka H, Shoyamac Y (2007) Improvement of artemisinin production by chitosan in hairy root cultures of *Artemisia annua* L. Biotechnol Lett 29:1143–1146
66. Ramani S, Jayabaskaran C (2008) Enhanced catharathine and vindoline production in suspension cultures of *Catharanthus roseus* by ultraviolet-B light. J Mol Signal 3:1–6
67. Rijhwani SK, Shanks JV (1998) Effect of elicitor dosage and exposure time on biosynthesis of indole alkaloids by *Catharanthus roseus* hairy root cultures. Biotechnol Prog 14:442–449
68. Romero-Puertas MC, McCarthy I, Sandalio LM, Palma JM, Corpas FJ, Gómez M, del Río LA (1999) Cadmium toxicity and oxidative metabolism of pea leaf peroxisomes. Free Rad Res 31(suppl):S25–S31
69. Rudrappa T, Neelwarne B, Kumar V, Lakshmanan V, Venkataramareddy SR, Aswathanarayana RG (2005) Peroxidase production from hairy root cultures of red beet (*Beta vulgaris*). Electron J Biotechnol 8:185–196
70. Satdive RK, Fulzele DP, Eapen S (2007) Enhanced production of azadirachtin by hairy root cultures of *Azadirachta indica* A. Juss by elicitation and media optimization. J Biotechnol 128:281–289
71. Shams-Ardakani M, Hemmati S, Mohagheghzadeh A (2005) Effect of elicitors on the enhancement of podophyllotoxin biosynthesis in suspension cultures of *Linum album*. DARU J Pharm Sci 13:56–60
72. Shi M, Kwok KW, Wu JY (2007) Enhancement of tanshinone production in *Salvia miltiorrhiza* Bunge (red or Chinese sage) hairy-root culture by hyperosmotic stress and yeast elicitor. Biotechnol Appl Biochem 46:191–196
73. Shilpa K, Varun K, Lakshmi BS (2010) An alternate method of natural drug production: eliciting plant secondary metabolite using plant cell culture. J Plant Sci 5:222–247
74. Shinde AN, Malpathak N, Fulzele DP (2009) Enhanced production of phytoestrogenic isoflavones from hairy root cultures of *Psoralea corylifolia* L. using elicitation and precursor feeding. Biotechnol Bioproc Eng 14:288–294
75. Singh G, Gavrieli J, Oakey JS, Curtis WR (1998) Interaction of methyl jasmonate, wounding and fungal elicitation during sesquiterpene induction in *Hyoscyamus muticus* in root cultures. Plant Cell Rep 17:391–395

76. Singh G, Reddy GR, Curtis WR (1994) Use of binding measurements to predict elicitor dosage requirements for secondary metabolite production from root cultures. Biotechnol Prog 10:365–371
77. Smetanska I (2008) Production of secondary metabolites using plant cell cultures. Adv Biochem Engin/Biotechnol 111:187–228
78. Stojakowska A, Burczyk J, Kisel W, Zych M, Banaś A, Duda T (2008) Effect of various elicitors on the accumulation and secretion of spiroketal enol ether diacetylenes in feverfew hariy root culture. Acta Soc Bot Pol 77:17–21
79. Syklowska-Baranek K, Pietrosiuk A, Kokoszka A, Furmanowa M (2009) Enhancement of taxane production in hairy root culture of *Taxus* × *media* var. Hicksii. J Plant Physiol 166:1950–1954
80. Talano MA, Agostini E, Medina MI, De Forchetti SM, Tigier HA (2003) Tomato (*Lycopersicon esculentum* cv. Pera) hairy root cultures: characterization and changes in peroxidase activity under NaCl treatment. In Vitro Cell Dev Biol-Plant 39:354–359
81. Thakore D, Srivastava AK, Sinha A (2012) Enhanced production of antihypertensive drug ajmalicine in transformed hairy root culture of *Catharanthus roseus* by application of stress factors in statistically optimized medium. Health, Energy and Environmental Perspectives. In: Srivastava MM, Khemani LD, Srivastava S (eds) Chemistry of Phytopotentials. Springer, Berlin, pp 39–42
82. Thimmaraju R, Bhagyalakshmi N, Narayan MS, Ravishankar GA (2003) Kinetics of pigment release from hairy root cultures of *Beta vulgaris* under the influence of pH, sonication, temperature and oxygen stress. Process Biochem 38:1069–1076
83. Toivonen L, Laakso S, Rosenqvist H (1992) The effect of temperature on hairy root cultures of *Catharanthus roseus*: growth, indole alkaloid accumulation and membrane lipid composition. Plant Cell Rep 11:395–399
84. Uozumi N, Kato Y, Nakashimada Y, Kabayashi T (1992) Excretion of peroxidase from horseradish hairy root in combination with ion supplementation. Appl Microb Biotechnol 37:560–565
85. Wang JW, Wu JY (2010) Tanshinone biosynthesis in *Salvia miltiorrhiza* and production in plant tissue cultures. Appl Microbiol Biotechnol 88:437–449
86. Wang JW, Zhang Z, Tan RX (2001) Stimulation of artemisinin production in *Artemisia annua* hairy roots by the elicitor from the endophytic *Colletotrichum* sp. Biotechnol Lett 23:857–860
87. Wang JW, Zheng LP, Tan RX (2006) The preparation of an elicitor from a fungal endophyte to enhance artemisinin production in hairy root cultures of *Artemisia annua* L. Chinese J Biotechnol 22:829–834 (in Chinese)
88. Wang JW, Zheng LP, Zhang B, Zou T (2009) Stimulation of artemisinin synthesis by combined cerebroside and nitric oxide elicitation in *Artemisia annua* hairy roots. Appl Microbiol Biotechnol 85:285–292
89. Wang YC, Zhang HX, Zhao B, Yuan XF (2001) Improved growth of *Artemisia annua* L hairy roots and artemisinin production under red light conditions. Biotechnol Lett 23:1971–1973
90. Wang Z-Y, Zhong J-J (2002) Repeated elicitation enhances taxane production in suspension cultures of *Taxus chinensis* in bioreactors. Biotechnol Lett 24:445–448
91. Weathers PJ, Towler MJ, Xu JF (2010) Bench to batch: advances in plant cell culture for producing useful products. Appl Microbiol Biotech 85:1339–1351
92. Wielanek M, Urbanek H (2006) Enhanced glucotropaeolin production in hairy root cultures of *Tropaeolum majus* L. by combining elicitation and precursor feeding. Plant Cell Tiss Organ Cult 86:177–186
93. Wu J, Zhong JJ (1999) Production of ginseng and its bioactive components in plant cell culture: current technological and applied aspects. J Biotechnol 68:89–99
94. Wu JY, Ng J, Shi M, Wu SJ (2007) Enhanced secondary metabolite (tanshinone) production of *Salvia miltiorrhiza* hairy roots in a novel root-bacteria coculture process. Appl Microbiol Biotechnol 77:543–550

95. Wu JY, Shi M (2008) Ultrahigh diterpenoid tanshinone production through repeated osmotic stress and elicitor stimulation in fed-batch culture of *Salvia miltiorrhiza* hairy roots. Appl Microbiol Biotechnol 78:441–448
96. Xu MJ, Dong JF, Zhu MY (2005) Effect of nitric oxide on catharanthine production and growth of *Catharanthus roseus* suspension cells. Biotechnol Bioeng 89:367–372
97. Xu MJ, Dong JF, Zhu MY (2005) Nitric oxide mediates the fungal elicitor-induced hypericin production of *Hypericum perforatum* cell suspension cultures through a jasmonic-acid-dependent signal pathway. Plant Physiol 139:991–998
98. Xu XJ, Hu XY, Neill SJ, Fang JY, Cai WM (2005) Fungal elicitor induces singlet oxygen generation, ethylene release and saponin synthesis in cultured cells of *Panax ginseng* C. A. Meyer. Plant Cell Physiol. 46:947–954
99. Yamazaki Y, Sudo H, Yamazaki M, Aimi N, Saito K (2003) Camptothecin biosynthetic genes in hairy roots of *Ophiorrhiza pumila*: cloning, characterization and differential expression in tissues and by stress compounds. Plant Cell Physiol 44:395–403
100. Yan Q, Hu ZD, Tan RX, Wu JY (2005) Efficient production and recovery of diterpenoid tanshinones in *Salvia miltiorrhiza* hairy root cultures with in situ adsorption, elicitation and semi-continuous operation. J Biotechnol 119:416–424
101. Yaoya S, Kanho H, Mikami Y, Itani T, Umehara K, Kuroyanagi M (2004) Umbelliferone released from hairy root cultures of *Pharbitis nil* treated with copper sulfate and its subsequent glucosylation. Biosci Biotechnol Biochem 68:1837–1841
102. Yu K-W, Murthy HN, Hahn E-J, Pa K-Y (2005) Ginsenoside production by hairy root cultures of *Panax ginseng*: influence of temperature and light quality. Biochem Eng J 23:53–56
103. Zhang B, Zheng LP, Wang JW (2012) Nitric oxide elicitation for secondary metabolite production in cultured plant cells. Appl Microbiol Biotechnol 93:455–466
104. Zhang B, Zou T, Yan Hua Lu YH, Wang JW (2010). Stimulation of artemisinin biosynthesis in Artemisia annua hairy roots by oligogalacturonides. Afri J Biotechnol 9: 3437-3442
105. Zhang C, Yan Q, Cheuk WK, Wu JY (2004) Enhancement of tanshinone production in *Salvia miltiorrhiza* hairy root culture by Ag^{+} elicitation and nutrient feeding. Planta Med 70:147–151
106. Zhang C-H, Wu J-Y (2003) Ethylene inhibitors enhance elicitor-induced paclitaxel production in suspension cultures of *Taxus* spp. Cells. Enzyme Microb Technol 32:71–77
107. Zhang H-C, Liu J-M, Lu H-Y, Gao S-L (2009) Enhanced flavonoid production in hairy root cultures of *Glycyrrhiza uralensis* Fisch by combining the over-expression of chalcone isomerase gene with the elicitation treatment. Plant Cell Rep 28:1205–1213
108. Zhang J, Xu M (2006) Effects of nitric oxide and methyl jasmonate on the baicalin production and cell growth in suspension cultures of *Scutellaria baicalensis*. Chin Sci Bull 23:374–379
109. Zhang L, Yang B, Lu BB, Kai GY, Wang ZN, Xia Y, Ding RX, Zhang HM, Sun XF, Chen WS, Tang KX (2007) Tropane alkaloids production in transgenic *Hyoscyamus niger* hairy root cultures over-expressing putrescine *N*-methyltransferase is methyl jasmonate-dependent. Planta 225:887–896
110. Zhao J, Davis LC, Verpoorte R (2005) Elicitor signal transduction leading to production of plant secondary metabolites. Biotechnol Adv 23:283–333
111. Zhao J, Hu Q, Guo Y-Q, Zhu W-H (2001) Elicitor-induced indole alkaloid biosynthesis in *Catharanthus roseus* cell cultures is related to Ca^{2+} influx and the oxidative burst. Plant Sci 161:423–431
112. Zhao JL, Zhou LG, Wu JY (2010) Effects of biotic and abiotic elicitors on cell growth and tanshinone accumulation in *Salvia miltiorrhiza* cell cultures. Appl Microbiol Biotechnol 87:137–144
113. Zhao JL, Zhou LG, Wu JY (2010) Promotion of *Salvia miltiorrhiza* hairy root growth and tanshinone production by polysaccharide–protein fractions of plant growth-promoting rhizobacterium *Bacillus cereus*. Process Biochem 45:1517–1522

114. Zheng LP, Guo YT, Wang JW, Tan RX (2008) Nitric oxide potentiates oligosaccharide-induced artemisinin production in *Artemisia annua* hairy roots. J Integr Plant Biol 50:49–55
115. Zheng LP, Zhang B, Zou T, Chen ZH, Wang JW (2010) Nitric oxide interacts with reactive oxygen species to regulate oligosaccharide-induced artemisinin biosynthesis in *Artemisia annua* hairy roots. J Med Plants Res 4:758–765
116. Zheng ZG, Wu M (2004) Cadmium treatment enhances the production of alkaloid secondary metabolites in *Catharanthus roseus*. Plant Sci 166:507–514
117. Zhong JJ, Seki T, Kinoshita S, Yoshida T (1991) Effect of light irradiation on anthocyanin production by suspended cultures of *Perilla frutescens*. Biotechnol Bioeng 38:653–658
118. Zhou L, Yang C, Li J, Wang S, Wu J (2003) Heptasaccharide and octasaccharide isolated from *Paris polyphylla* var. *yunnanensis* and their plant growth regulatory activity. Plant Sci 165:571–575
119. Zhou LG, Cao XD, Zhang RF, Peng YL, Zhao SJ, Wu JY (2007) Stimulation of saponin production in *Panax ginseng* hairy roots by two oligosaccarides from *Paris polyphylla* var. *yunnanensis*. Biotechnol Lett 29:631–634
120. Zhou LG, Wu JY (2006) Development and application of medicinal plant tissue cultures for production of drugs and herbal medicinals in China. Nat Prod Rep 23:789–810
121. Zhu L, Cullen WR (1995) Effects of some heavy metals on cell suspension cultures of *Catharanthus roseus*. J Environ Sci 7:60–65
122. Zimmerli L, Métraux J-P, Mauch-Mani B (2001) β-aminobutyric acid-induced protection of *Arabidopsis* against the necrotrophic fungus *Botrytis cinerea*. Plant Physiol 126:517–523

Adv Biochem Eng Biotechnol (2013) 134: 91–114
DOI: 10.1007/10_2013_181

Published Online: 19 April 2013

Hairy Root Culture: Bioreactor Design and Process Intensification

Amanda R. Stiles and Chun-Zhao Liu

Abstract The cultivation of hairy roots for the production of secondary metabolites offers numerous advantages; hairy roots have a fast growth rate, are genetically stable, and are relatively simple to maintain in phytohormone free media. Hairy roots provide a continuous source of secondary metabolites, and are useful for the production of chemicals for pharmaceuticals, cosmetics, and food additives. In order for hairy roots to be utilized on a commercial scale, it is necessary to scale-up their production. Over the last several decades, significant research has been conducted on the cultivation of hairy roots in various types of bioreactor systems. In this review, we discuss the advantages and disadvantages of various bioreactor systems, the major factors related to large-scale bioreactor cultures, process intensification technologies and overview the mathematical models and computer-aided methods that have been utilized for bioreactor design and development.

Keywords Bioreactor · Computational fluid dynamics · Hairy roots · Process intensification

Contents

A. R. Stiles · C.-Z. Liu (✉)
National Key Laboratory of Biochemical Engineering, Institute of Process Engineering, Chinese Academy of Sciences, Beijing 100190, People's Republic of China
e-mail: czliu@home.ipe.ac.cn

1 Introduction

Medicinal plants have been used worldwide for thousands of years. Even today, 80 percent of the world's population depends on plant derived medicines for their daily health [1]. Thousands of plant species are used for treatments and therapies. Plant secondary metabolites, which include phenolic compounds, alkaloids, flavonoids, and polysaccharides among others, are the main components for disease treatment. In recent years, increasing numbers of medicinal plants have been discovered for treating infections [2, 3], cardiovascular ailments [4, 5], cancers [6, 7], and even AIDS [8]. In addition, transgenic plants have the potential to produce pharmaceutical proteins, which would have a significant impact on the pharmaceutical industry [9, 10].

However, many medicinal plants are difficult to cultivate and must be collected from the wild, which puts them at risk of becoming threatened or endangered [11]. *In vitro* cultivation offers an efficient method for the large-scale production of plants for medicinal purposes, thereby helping to safeguard critically endangered species. Various in vitro methods are utilized for controllable plant sources for the commercial production of secondary metabolites, including cell culture, callus cultures, organogenesis, embryogenesis and root cultures. The large-scale production of cell, tissue, and organ culture using bioreactors has been accomplished in many species and shows potential for industrial production of naturally occurring as well as novel metabolites.

The cultivation of hairy roots is an in vitro method that has many advantages for commercial production. Hairy roots are developed by infecting explants with *Agrobacterium rhizogenes,* a soil bacterium that integrates a DNA segment into the host plant genome which results in the active proliferation of the roots [12]. Compared to normal root cultures, hairy roots are fast-growing, genetically and biochemically stable, easy to maintain, and are able to grow in phytohormone free media. They offer a continuous source of secondary metabolites, and they are garnering increasing interest for the production of chemicals for pharmaceuticals, cosmetics, and food additives. They are especially useful for the production of secondary metabolites as many products are synthesized in roots but not formed in suspension or callus culture [13]. In addition, due to their stable nature, hairy roots serve as an excellent model system to study plant metabolism and physiology [14]. In addition, they also serve as a promising transgenic system; transformed roots of many plant species have been widely studied for the in vitro production of secondary metabolites [15–18]. Genetic manipulation can increase the biosynthetic capacity [19, 20], produce multiple secondary compounds, and serve as a heterologous system for compound production, including foreign proteins [21]. Once a transgenic line is produced, a single hair root from the explant tissue serves as a clone [22], and once generated, they are capable of unlimited propagation [23]. Although the majority of research on hairy roots has been conducted in shake flasks [24], production in bioreactors allows for controlled conditions that minimize variations in the yield and quality of the product, and bioreactors allow for the optimization of conditions for increased cell growth and secondary metabolite production [25]. Optimization of bioreactor systems for the cultivation of hairy roots is necessary for their scale-up production for industrial applications.

The cultivation of hairy roots in a bioreactor system creates unique issues, hairy roots are morphologically different among plant species; characteristics such as the thickness, length, hairiness and branching of the roots are affected by both the plant species and the *Agrobacterium* strain used for the induction of hairy roots [25]. In addition, both cell growth and metabolite production are non-homogeneous in hairy roots, further complicating bioreactor optimization [21]. In this review, we discuss the advantages and disadvantages of various bioreactor systems, discuss the major factors related to large-scale bioreactor cultures, discuss process intensification technologies that have been applied in hairy root bioreactors for industrial production, and overview the mathematical models and computer-aided methods that have been utilized for bioreactor design and development.

2 Bioreactor Types

Bioreactors are generally self-contained, sterile environments that include liquid nutrients, inflow and outflow systems for liquid and air, and are designed to optimize and monitor culture conditions. In general, they provide the ability to control micro-environmental conditions, such as pH, dissolved gases, aeration, and temperature [26]. Bioreactors used to culture hairy roots can be broadly divided into either liquid-phase or gas-phase reactors. Briefly, in liquid-phase reactors, roots are submerged in the medium; therefore, they are also sometimes termed 'submerged reactors' while in gas-phase reactors, the roots are exposed to air or other gas mixtures, and nutrients are generally delivered to the roots as droplets of various sizes [24]. Additional types include hybrid reactors, which are a combination of liquid and gas-phase reactors, and disposable reactors. Schematic diagrams of several of the different types of hairy root bioreactors are included in a review by Mishra and Ranjan [13]. There are advantages and disadvantages to the many types of bioreactors that have been successfully developed for cultivating hairy roots; therefore, there are many factors involved in selecting the best bioreactor design.An overview of the different bioreactor types is included in Table 1.

Table 1 Advantages and disadvantages of major types of bioreactors

Bioreactor types		Advantages	Disadvantages
Liquid-phase reactors	Stirred tank	Mixing and breaking up air bubbles prevent cell aggregation and enhance oxygenation	High shear force, complicated configuration, increased exposure to contaminants, high energy consumption, difficult to optimize multiple variables
	Airlift and Bubble column	Low shear stress, simple design and construction, low contamination, low maintenance	Foaming induced by large air volumes, non-uniform growth of roots within the reactor (roots can 'float' to the top)
Gas-phase reactors	Nutrient mist	Abundant oxygen supply, low sugar concentration and high space utilization	Complex construction, high energy consumption, labor intensive set-up
	Trickle-bed	Abundant oxygen supply and low energy consumption	May produce a viscous liquid film on the roots creating a high mass transfer barrier, labor intensive set-up

2.1 Liquid-Phase Reactors

In liquid-phase reactors, the culture space is filled with liquid medium and various methods are employed to provide aeration. Because the roots are submerged in liquid medium, mixing and mass transfer become the main bottlenecks in the design and scale-up. Due to the low solubility of gases in the liquid phase, gas exchange limitations and insufficient nutrient transport have been reported [25]. A study by Curtis [27] suggested that at lower tissue concentrations (<10 g/L DW), hairy roots can be grown in virtually any type of liquid-phase reactor, but at higher concentrations (>10 g/L DW), they are likely to encounter scale-up limitations [27]. Design considerations include mechanisms to provide sufficient aeration, nutrient mixing, and immobilization of the roots. The differences in mixing and aeration represent the main design differences among the various types of liquid-phase reactors and several of the most common designs are described below, including stirred tank and pneumatic reactors. Methods for immobilizing hairy roots include meshes (horizontal or vertical), cages, and polyurethane foam [25].

Early studies examined the use of stirred tank reactors, a design commonly used with cell suspension cultures, for the cultivation of hairy roots, including *Calystegia sepium* and *Atropa belladonna* hairy roots for the production of tropane alkaloids [28]. However, the use of the impeller in the traditional stirred tank reactors damaged the sensitive plant tissues, resulting in callus formation and poor biomass production [29]. Therefore, alternate methods to improve the supply of oxygen were tested, and several studies focused on determining ways to improve impeller performance by modifying the internal reactor [30–32]. Stirred tank reactors demonstrated varied success, *Catharanthus trichophyllus* cultivated in both shake flasks and a stirred tank reactor showed a similar alkaloid composition [33], while *Panax ginseng* hairy roots cultivated in a stirred bioreactor, achieved growth approximately threefold higher than in shake flasks [34]. More recent studies have examined the use of modified stirred tank reactors. Srivastava and Srivastava et al. [35], developed a stirred tank reactor in which the roots were anchored onto a polyurethane foam disk with a low shear impeller for mixing and nutrient transfer placed below the disk. This method allowed for sufficient mixing while protecting the roots from shear stress, although *Azadirachta indica* hairy roots were still unable to produce the same levels of biomass or azadirachtin concentrations obtained via shake flask cultivation [36].

Pneumatic bioreactors, which include both bubble column and airlift reactors, consist of a cylinder with a sparger at the bottom that releases an air or a gas mixture through the solution for both aeration and mixing. Bubble column bioreactors are one of the simplest and easiest types to scale-up. The use of bubbles minimizes shear stress on the cultures, however, the bubbling must be increased as the hairy root biomass increases [37]. Bubble column reactors require low capital, have low operational costs, and contain no moving mechanical parts [21]. However, due to the undefined flow pattern, the liquid is not uniformly mixed [37], and in high density cultures, bubble column reactors often result in reduced growth performance [38]. In high biomass cultures, the bubbles may coalesce, reducing the gas–liquid interface area [21]. The introduction of multiple spargers to deliver oxygen into areas with a high cell density may improve biomass; gas introduced at multiple sections in a multi-compartment bubble column reactor resulted in a high biomass density (in some sections up to 17 g/L), but still provided poor bulk mixing [38]. The use of a microporous polypropylene membrane tubing, in addition to the sparger (required for bulk mixing),to provide supplementary oxygen directly to the root bed, resulted in a biomass increase of 32 % compared to the use of the sparger alone [39]. Various operating parameters may be optimized to improve the biomass and secondary metabolite production in bubble column reactors; for example, the production of alkaloids *by S. parviflora* hairy roots in a bubble column reactor was increased by optimizing the culture period, initial inoculum density, and aeration rate [40].

In contrast to bubble column reactors, airlift bioreactors contain a draft tube (either internal or external) to prevent bubbles from coalescing, and to enhance oxygen mass transfer by increasing the number of bubbles. Airlift reactors reduce shear stress, distribute shear stress more evenly, consume little energy, and promote a cylindrical mixing of the medium. Airlift bioreactors have been used extensively used for hairy roots since the initiation of hairy root bioreactor studies, for species including *Panax ginseng* [41], *Armoracia rusticana* [42], *Trigonella foenumgraceum* [43], *Lippia dulcis* [44], *Lithospermum erythrorhizon* [45], *Ophiorrhiza pumila* [46], and *Echinacea purpurea* [47]. In some cases, airlift bioreactors have been highly effective compared to shake flask cultures; *Pueraria phaseoloides* hairy roots cultivated in a 2.5 L airlift bioreactor produced 200 times as much puerarin as in a 250 ml flask culture [48]. The growth of *Panex ginseng* hairy roots inoculated into a 5 L airlift bioreactor increased by approximately 55-fold after 39 days in a 5 L airlift bioreactor, and 38-fold after 40 days in a 19 L airlift bioreactor [34]. In airlift reactors, both the liquid and the aeration are driven by externally supplied air; however, similar to bubble column reactors, they are generally not suitable for high-density cultures because they result in insufficient mixing and oxygen mass transfer [37].

2.2 Gas-Phase Reactors

In gas-phase reactors, the roots are exposed to either an air or a gas mixture and the nutrients are delivered as droplets of different sizes. They have been widely utilized in plant tissue culture, and are useful for hairy root cultures because they provide an abundant oxygen supply. In gas-phase reactors, the deposition of nutrients is the key step in nutrient mass transfer; however while some deposition is required, excessive deposition will result in a thick liquid layer on the root surface that inhibits gas transfer [49]. In addition, if drainage is insufficient in gas-phase reactors, the root bed may retain liquid and further inhibit gas transfer [50]. In contrast to liquid-phase reactors, in which the root hairs increase the flow resistance of the media, in gas-phase reactors, the root hairs are believed to be beneficial for growth because they enhance mist capture and may improve reactor performance [51]. Although gas-phase reactors result in improved oxygen transfer, they still require a matrix for anchoring the hairy roots (such as mesh trays, stainless steel mesh cylinders), and the requirement for uniform loading may be labor intensive [25]. Examples of gas-phase reactors include: nutrient mist, nutrient sprinkle, and trickle bed reactors.

In mist bioreactors, the nutrient mist may be produced by ultrasonic methods, nozzles, or compressed air. The mist dispersed in the bioreactor has a large specific surface area that leads to a high oxygen transfer rate from gas into medium, and may account for the higher biomass production in mist bioreactors compared with other types. Compared with both shake flasks and bubble column reactors, mist reactors are less oxygen-limited; this was demonstrated by an analysis of the expression of alcohol dehydrogenase, an indicator of oxygen stress, in hairy root cultures. In a mist bioreactor there was no detectable expression up to a packing density of approximately 37 % (v/v) while in contrast, there was a significant expression of alcohol dehydrogenase in roots grown in bubble column bioreactors at densities of only 6 % [52]. Parameters that may be adjusted in a mist reactor to optimize nutrient delivery include the misting cycle time and the medium flow rate. A mist bioreactor scaled up from 1–20 L for *A. hypogaea* hairy root cultures demonstrated that increasing the misting cycles longer that 2–3 min was detrimental, but increasing the medium flow rate during the exponential growth phase increased the growth rates and biomass yields [53].

Nutrient sprinkle reactors produce larger droplets than mist reactors. Kochen et al. [54] developed and tested a nutrient sprinkle reactor for the production of ginsenosodies from *Panax quinquifolium*; although the biomass increase in the nutrient sprinkle reactor was slightly lower than shake flask cultures, the ginsenoside content was doubled when the cultivation of both *Salvia officinalis* hairy roots and shoot cultures was examined in a nutrient sprinkle bioreactor; the hairy roots achieved an 18-fold increase in biomass after 40 days of culture and the production of rosmarinic acid was approximately 1.3-fold higher in the hairy roots compared to shoot cultures. Overall, the reactor was considered unsuitable for the cultivation of shoots due to hyperhydricity issues [55]. *Salvia sclarea* hairy root cultures cultured in a nutrient sprinkle reactor produced 9 and 3.8 times as much aethiopinone and salvipisone, respectively, as roots cultured in shake flasks [56].

Trickle-bed bioreactors produce the largest sized droplets compared with mist and nutrient sprinkle bioreactors; this results in the formation of a thicker liquid film on the plant tissue surface and creates a disadvantage for gas transfer [24]. In addition, the tendency for the root bed to accumulate liquid and the absence of agitation are major limitations in the large-scale design of trickle-bed bioreactors [57]. However, they supply a larger volume of liquid nutrients to the cultures and use less energy compared with mist bioreactors. Ramakrishnan et al. [58] scaled-up a 14 L trickle-bed bioreactor for *Hyoscyamus muticus* hairy root culture, and their analysis of the mass transfer and fluid dynamic characteristics provided a method for the design in the scale-up of trickle-bed bioreactors [58].

Several comparison studies have been conducted to determine the optimal bioreactor type for a particular species of hairy roots. The production of *Artemisia annua* hairy roots was compared in mist and bubble bioreactors, and while the artemisinin content of *Artemisia annua* hairy roots was higher in the mist reactor [59], the overall biomass was higher in the bubble column reactor [60]. A comparison of the growth and productivity of transgenic tobacco hairy roots in a mist bioreactor and an airlift bioreactor showed that the synthesis of murine interleukin 12 was increased by approximately 50 % in the mist bioreactor [9]. *Tagetes patula* hairy root cultures cultivated in a bubble bioreactor, a sprinkle bioreactor, and a mist bioreactor achieved the highest growth rate, biomass production, and metabolite production in the mist bioreactor [61]. Srivastava and Srivastava [36] examined azadirachtin production by *Azadirachta indica* hairy roots in a stirred tank reactor, bubble column reactor, nutrient spray reactor, and nutrient mist reactor and found that the nutrient mist reactor was able to produce biomass and azadirachtin levels most similar to those achieved in optimized shake flask cultures [35].

2.3 Novel Bioreactors

Although overall, gas-phase reactors have the advantage of improved oxygen transfer efficiency, a disadvantage of gas-phase reactors is the necessity for the manual distribution of the roots in the growth chamber, a labor intensive process. One method to address this issue was the development of a hybrid reactor, a reactor in which the initial phase of the reactor was liquid-phase to allow the roots to attach uniformly to the anchoring system, and once attached, the reactor was switched to a gas-phase reactor. A large-scale (400L) reactor was developed in the late 90's using this method (a combination of a bubble column and spray reactor) for *Datura stramonium* hairy roots [62], and Ramakrishnan et al. [50] also developed a hybrid reactor in which the reactor was initially run as a bubble column, and then switched to a gas-phase reactor for the improved efficiency.

Disposable bioreactors are another alternative to the traditional culture systems. Disposable reactors utilize sterile plastic chambers that may be inoculated and then discarded once the roots are harvested, thus lowering operation costs by eliminating the need for cleaning or sterilization [63]. *Hyoscyamus muticus* and *Panax ginseng* hairy roots were cultivated in wave bioreactors, disposable bioreactors consisting of screw-cap sealed plastic bags under laminar fluid flow conditions. Under optimal culture conditions, the *Hyoscyamus muticus* root biomass increased by approximately 120-fold after 28 days, and the *Panax ginseng* biomass increased by 28-fold (compared to only 12.1-fold in a spray reactor) [63].

3 Bioreactor Parameters

Secondary metabolite biosynthesis in hairy roots is genetically controlled, but is influenced by nutritional and environmental factors. In addition, the rheological properties of hairy root cultures vary from one species to another and sometimes even within clones of the same species [13]. Multiple factors must be considered for the successful scale-up of bioreactor systems, including the physiology, morphology, and stress sensitivity of hairy roots. Submerged hairy roots tend to form dense clumps due to their tendency to form lateral branches and the hairiness of the roots [64, 65]. In liquid-phase reactors, these dense clumps resist fluid flow and therefore the delivery of nutrients to the tissues, creating one of the major challenges in reactor scale-up.The main bottleneck in scale-up, is the delivery of sufficient oxygen, but other factors, such as light, other gases, temperature, nutrients, and inoculum conditions also affect the growth and productivity of hairy root cultures [37].

3.1 Oxygen and Other Gases

Dissolved oxygen plays an important role in the bioreactor microenvironment, though oxygen requirements may vary from species to species. Hairy roots cultured in bioreactors have a tendency to form dense root clumps that limit oxygen transfer; therefore, understanding oxygen transfer and supply has been one of the most highly studied research topics in hairy root bioreactor design. In general, it is essential that the dissolved oxygen concentration remains above a critical level at all the times for optimal cell growth [66]. Shiao and Doran found that as root hairiness increased, the oxygen mass transfer boundary layer increased, limiting oxygen consumption in hairy root cultures [67]. A mathematical model developed to estimate the oxygen mass transfer coefficient in bubble column reactors demonstrated that due to the low solubility of oxygen and the high oxygen consumption of hairy roots, the oxygen concentration near hairy roots was significantly different from the bulk fluid [68]. When dissolved oxygen drops below a certain point, respiration is impaired, fermentation begins, and toxic byproducts are produced.

In liquid-phase reactors, the oxygen concentration is generally measured using a sterile electrode, and can be regulated by agitation, aeration, gas flow, and bubble size. Analyzing the transport of oxygen requires both characterizing the oxygen gradients and the thermodynamic equilibria that determine the solubility of oxygen in the media and plant tissue [69]. The oxygen concentrations in liquid cultures depend on the presence of oxygen in the gas phase above the medium, in the air bubbles inside the medium, and on the dissolved oxygen in the medium itself. Oxygen transport in hairy root cultures in a liquid-phase reactor is mainly influenced by the mass transfer resistance at the liquid–solid phase, rather than gas–liquid transfer [23]. Early studies using high liquid velocities to remove the hydrodynamic boundary layers, demonstrated that culture growth did not continue exponentially and indicated that the mucilage and root hairs are a barrier to oxygen transfer [57]. Other studies corroborated that the root hairs contributed to a pressure drop in convective flow reactors [70, 71]. Overall, due to the morphology of hairy roots, it is unlikely that it will be possible to completely eradicate oxygen limitations in liquid-phase reactors. It is possible to optimize and improve growth, using methods such as increasing the aeration rate; increasing the aeration rate from 0.002 m^3/h to 0.012 m^3/h improved cichoric acid production of *Echinacea purpurea* cultures cultivated in a modified air lift reactor [47].

Gas-phase reactors have the advantage on greater oxygen transfer and reduce the accumulation of mucilage; however, the thick roots can trap liquid, thereby reducing the contact of the roots with the gas phase. Williams and Doran [57] have suggested this could be overcome with the use of mechanical or pneumatic agitation [57], and methods to reduce cycle on time and study drainage characteristics have been tested to reduce this issue. For example, Ramakrishnan et al. found that respiration in hairy roots increased sharply with increased liquid spraying rate at relatively high tissue concentration in trickle-bed reactor [50].

Many methods have been tested to improve oxygen availability for hairy roots in bioreactors. In liquid-phase reactors, additional tubing has been added to improve oxygen availability, such as Kanokwaree et al. added microporous polypropylene tubing to bubble bioreactors containing *Atropa belladonna* hairy root cultures, thereby diminishing the oxygen limitation and improving both biomass and the accumulation of metabolites [39]. Kino-Oka et al. [72] found that increasing the medium flow rate in a bubble column reactor enhanced the dissolved oxygen concentration but simultaneously increased shear stress. To overcome this, a radial flow reactor was constructed that successfully released the oxygen transfer barrier in high-density red beet hairy root cultures [72]. A transgenic method was also tested to enhance growth under low oxygen conditions; transgenic *Arabidopsis thaliana* L hairy root lines that contained alcohol dehydrogenase and pyruvate decarboxylase were generated. The transformed root lines showed a similar growth rate under low oxygen conditions as those under full aeration [73].

The biomass and secondary metabolite production is affected by the gas composition in the media. The gas composition within a bioreactor is impacted by its volume, the volume of the media, and the bioreactor design. In addition to the oxygen concentrations within a bioreactor, carbon dioxide and ethylene also play a role. Carbon dioxide has the potential, in some species, to improve root growth. For example, beet hairy roots cultured in a nutrient mist reactor enriched with 1 % carbon dioxide showed a 15 % increase in biomass compared with tissue cultured in ambient air [74]. This response was due to the fact that it reduced the lag phase of the culture, allowing the culture to accumulate biomass faster, not because it increased the actual growth rate. However, the effect of carbon dioxide may vary dependent on the species; *A. annua* hairy roots grown in enriched carbon dioxide appeared healthier, but did not grow faster [75]. Sung and Huang [76] examined the effect of ethylene on the growth and secondary metabolite production of *Stizolobium hassjoo* hairy roots. The accumulation of ethylene in the headspace resulted in a reduction in both biomass and secondary metabolite production (L-DOPA), and when an ethylene inhibitor was added, both biomass and L-DOPA increased [76].

3.2 Light and Temperature

Light plays a role in both growth and secondary metabolite production, and due to the morphology of the hairy root cultures, it is difficult to achieve uniform illumination. The stimulatory effect of light on the formation of secondary compounds has been demonstrated in plant species including *Perilla fruitescnens* and *Artemisia annua* [77–79], and *Ipomea aquatica* hairy roots grown in the light produced twice the biomass and four times the peroxidase as roots grown in the dark [80]. Many hairy root lines have been documented to change colors, to colors such as green or purple when exposed to light [78, 82]. In general, the green color is due to the development of chloroplasts that are fully capable of photosynthesis and the metabolic capabilities of the green roots are distinct from their non-green counterparts [83]. *Echinacea purpurea* hairy root cultures exposed to light develope a dark purple color which is related to anthocyanin production [77]. Abbasi et al. [77] studied the effect of light on the accumulation of caffeic acid derivatives (CADs) in *Echinacea purpurea* hairy root cultures and found that all of the CADs were significantly different between light-and dark-grown roots. Bhadra et al. [81] studied the effect of light on indole alkaloid accumulation by *Catharanthus roseus* hairy root cultures, and found the alkaloid total yield was also significantly different between light- and dark-grown hairy roots, and that the production of certain alkaloids during specific growth phases shifted when light conditions were altered [81]. Hairy roots that do not change colors in response to light may also show profound alterations in secondary metabolism in response to light [24].

The photoperiod and light intensity are interrelated, and both are important to hairy root cultivation. For example, Liu et al. [84] found that 16/8 h light/dark was optimum for artemisinin production by *Artemisia annua* hairy root cultures, and Wang et al. [78] found that red light had the greatest impact compared to all other light regimes. A comparison of *Panax ginseng* hairy root cultures under several different light regimes (dark, fluorescent, metal halide, monochromatic red, monochromatic blue, and red plus blue) showed that the ginsenoside production was highest in cultures grown under fluorescent light, while the biomass was higher in cultures grown under either red light or in the dark [85]. Temperature also plays an important role and may vary from species to species. A study on the effects of temperature on *Solanum aviculare* hairy roots identified 25 °C as the optimal temperature [86], while Hilton and Rhodes [87] found that *Datura stramonium* hairy roots cultivated at 30 °C had a biomass approximately 4-fold greater, and the production of hyoscyamine was up to 7-fold higher. In *Panax ginseng* hairy root cultures, a comparison of different temperatures demonstrated that the biomass of the hairy roots was highest in cultures incubated at 20 °C/13 °C for 16/8 h day cycles, while the production of total ginsenosides was optimum at cultures incubated at 25 °C/25 °C [85].

3.3 Nutrient Medium Composition

In experiments with *Panax ginseng* hairy roots, Sivakumar et al. [88] found that the commonly utilized MS medium resulted in low hairy root growth, biomass, and ginsenoside content, indicating that the nutrient composition could be optimized. They identified nutrient availability as the major chemical factor for scale-up and suggested that mineral elements are an important regulatory factor in hairy root growth and biomass production [88]. In addition, periodic estimations of specific nutrients at different periods in bioreactors can provide information regarding nutrient uptake, biomass, and metabolic production. Generally, the medium composition is modified with respect to its concentration of carbon, nitrogen, phosphorus [38, 89], and macronutrients [88]. A factorial experiment using *Artemisia annua* hairy roots was employed to determine the relationship between four factors (phosphate, nitrate, sucrose, and culture inoculum age) at three levels; results indicated there was a significant interaction between sucrose and nitrate although the other factors were less interactive [90]. Depending on the relationship between the biosynthesis of the secondary metabolites and biomass production, the media can either optimized for both growth and secondary metabolite production simultaneously or a two-stage strategy may be developed. A study evaluating the production of tropane alkaloids by *Datura stramonium* L. hairy roots used a factorial design and mathematical modeling approach to evaluate the relationship between nutritional status and elicitation [91]. They concluded that simulation with their developed model allowed them to predict the response to elicitation in response to nitrate and calcium levels, and that if hairy roots were nitrate limited; they showed a lowered response to jasmonic acid elicitation. A study on the effect of MS medium dilution on the biomass and accumulation of phenols and flavonoids in *Hypericum perforatum* adventitious roots found that the peak root growth was at four weeks using half-strength MS, and the content of total phenols and flavonoids were also highest using the quarter or half-strength MS [92].

4 Process Intensification

Process intensification approaches are utilized in plant cell and tissue cultures to enhance biomass production and metabolite yield. Elicitation, the introduction of compounds that trigger an increase in metabolite production, and ultrasound, a method utilized to enhance mass transfer in plant cell cultures, have both been applied to hairy root bioreactor cultures.

4.1 Elicitation

Elicitors are compounds that trigger the increased production of compounds in plant cells such as pigments, flavones, phytoalexins, and other chemicals related to defense [93]. In general, elicitors are classified as either biotic or abiotic. Biotic elicitors are derived from a pathogen or from the plant itself [94], while abiotic elicitors are either physical or chemical compounds. The effectiveness of elicitation depends on multiple variables, such as the specificity of the elicitor, the concentration, treatment interval, culture growth stage, medium composition, and light.

Abiotic elicitors are generally less expensive than biotic, and biotic elicitors also have the disadvantage that they may require additional facilities for the cultivation of the source microorganism [95]. In many studies, both abiotic and biotic elicitors are tested. For example, Pitta-Alvarez et al. [96] tested the effects of different biotic (salicylic acid, yeast extract) and abiotic ($CaCl_2$, $AgNO_3$, $CdCl_2$) elicitors on the accumulation and release of scopolamine and hyoscyamine in *Brigmansia candida* hairy root cultures [96]. Salicylic acid significantly increased the release of both alkaloids, $AgNO_3$ and yeast extract preferentially increased the release of scopolamine, and $CdCl_2$ increased the release of both alkaloids but also inhibited growth. *Azadirachta indica* A. Juss hairy roots exposed to a biotic elicitor (the fungus *Claviceps purpurea*), 100 mM jasmonic acid, or 200 mM salicylic acid achieved an approximate 5, 6, and 9-fold increase in azadirachtin yield, respectively [97]. A review by Georgiev [95] includes additional examples of secondary metabolite yield increases in response to elicitation of hairy roots [95].

Methyl jasmonate is a potent elicitor due to its ability to induce plant signal transduction pathways; *Salvia sclarea* hairy roots grown in a sprinkle bioreactor showed a 9-fold increase in aethiopione and a 3.8-fold increase in salvipisone when they were exposed to 125 μM methyl jasmonate for 7 days [56]. The application of 100 μM methyl jasmonate applied to *Silybum marianum* hairy roots cultured in a 2 L airlift bioreactor, resulted in a 1.6-fold increase in silymarin accumulation after only 72 h [98], and *Artemisia annua* L. hairy roots treated with 150 mg chitosan/L, 200 μM methyl jasmonate, or 2 mg/mL yeast extract resulted in a 6-fold, 3-fold, and 5-fold increase in artemisinin production, respectively [99].

Although elicitors may increase metabolite production, they may also simultaneously reduce biomass. *Beta vulgaris* hairy roots exposed to biotic elicitors including purified microbial glycans (200-500 mg/L), whole microbial culture extract (0.25–1.25 %), and culture filtrates (5–25 %, v/v) achieved significantly increased betalain production in the *Penicilium notatum* DCP-treated cultures, but all elicitors resulted in a reduction of biomass, reducing overall yields [100]. However, this is not always the case; vanadyl sulfate ($VOSO_4$), an abiotic elicitor, resulted in an 8-fold increase in biomass compared to the control in *Ambrosia artemisiifolia* hairy root cultures when 50 mg/L was applied for 72 h to 16-day-old cultures [101]. Due to the fact that in some cases elicitation inhibits root growth, a two-stage culture strategy, in which elicitation is conducted once growth is complete, may be used in scaled-up processes. For example, in elicitation studies with *Beta vulgaris* hairy roots cultured in a bubble-column reactor, the elicitor was added at late exponential phase, resulting in approximately a 47 % increase in betalain production compared to controls [100].

Elicitors may also increase oxidative stress; *Silybum marianum* hairy roots exposed to 100 uM methyl jasmonate for 72 h showed the increased expression of antioxidant enzymes such as ascorbate peroxidase and guanicol peroxidase by 1.3 and 3.2-fold, respectively [98]. Although our current understanding of the mode of action of elicitors is related almost exclusively to secondary metabolism, primary metabolism may also be affected. Vasconsuelo and Boland [94] suggest that elicitation may also affect processes that regulate levels of secondary metabolites, such as vacuolar transport and compound storage, and that increasing our overall understanding these processes will be helpful to improve secondary metabolite production [94].

4.2 Ultrasound

Ultrasound has been widely applied in chemistry, chemical engineering, medicine, and biology for its chemical, physical and biological effects. In plant cell cultures, ultrasonic treatments have been employed to enhance metabolism and mass transfer, examples include, *Lithospermum erythrorhizon*,*Panax ginseng*, and *Taxus chinensis* [102–104]. The effective distance of the ultrasound wave is short due to the attenuation of ultrasound wave [105, 106]; however, ultrasound has been successfully employed in hairy root cultures to intensify mass transfer in order to stimulate both root growth and secondary metabolite production. Cichoric acid production was increased in *Echinaciea purpurea* hairy root cultures by the use of six minute ultrasound treatments in a 2 L bioreactor [47]. The acoustic model has been utilized to understand the effects of ultrasound treatments on fluid flow [107, 108]; however, an optimal design for the scale-up of a sonobioreactor that would provide uniform ultrasound has not yet been developed.

5 Bioreactor Design and Scale-Up

5.1 Mathematical Models

Root tissues are very different from the common microbial or cell cultures, and bioreactors for hairy root culture are more difficult to operate, control, and scale-up. The high number of variables results in complicated methods of operation. Developing growth models for bioreactor cultures is part of the necessary research for the design, operation, and scale-up for the industrial production of hairy roots. The use of mathematical models aids in the design of reactors to improve biomass, secondary metabolite production, and secondary metabolite recovery.

Several methods have been established to monitor the process of plant cell growth, Kim et al. [109] developed a growth model using *Tagetes erecta* T3 hairy roots as a model system [109]. In this study, a mathematical model of branching dynamics was constructed to simulate the branching rules for hairy roots, and the growth rate and biomass were monitored using the population balance equation and the Monod equation. This model functioned to illustrate the root growing process and aid in bioreactor scale-up. More recently, a medium-throughput automatic image recognition system was developed to provide quantitative data regarding the growth patterns and secondary metabolite production (the red betacyanin pigment) of *B. vulgaris* hairy roots. In this system, they quantified three main growth processes; tip elongation, branching and secondary thickening, to evaluate growth, and quantified the red color produced by betacyanin to evaluate secondary metabolite production. This type of approach helps to provide quantitative data that may be used for the creation of a structured growth model for hairy roots via mathematical modeling [110].

In both shake flasks and bioreactors, hairy roots tend to form a dense network of tissue with younger, more lightly packed roots surrounding the older core. Bastian et al. [111] developed a general growth model that included hairy root elongation, branching, secondary growth, and nutrient transport functions in dense root networks. After certain simplifications, the model was simulated with the experiment data using *Ophiorrhiza mungos* hairy root culture. This model could be used to describe the uptake kinetics and biomass increase of *O. mungos* hairy roots grown in shake flasks, and helped with the optimization of process parameters in hairy root culture. Oxygen transfer is a critical factor in bioreactor cultures because in high density cultures; the oxygen uptake rate is generally faster than the oxygen mass transfer rate. Early experiments determined that oxygen consumption within the hairy root mass was restricted by the rate of convective mass transfer within the mass, and increasing the flow rate around the mass would induce higher rates of oxygen consumption [112].

The growth and metabolism of plant cells are distinct when they are cultivated in different environments; therefore, in order to achieve maximum production, it is necessary to monitor the cell and tissue growth, nutrient consumption, mass transfer, and rheology. Ramakrishnan and Curtis [58] detailed the oxygen transfer process in a plant tissue culture system using the uptake equilibrium and diffusion equilibrium. The mass transport process was divided into several steps including transport in the gas phase, transfer from gas into liquid phase, and transport from liquid to tissue. They constructed models of oxygen transport and oxygen consumption to estimate the solid–liquid mass transfer coefficient in hairy root cultures in a trickle bed bioreactor, which could then be used for the design of scaled-up reactors [58]. The mathematical analysis of oxygen transport into the plant tissue culture provided a quantitative understanding of the culture conditions; this is helpful for understanding oxygen transport limitations, which could then guide bioreactor design. Recently, a mathematical model was developed to evaluate and optimize operating parameters such as the mist flow rate, the nutrient concentration, the timing for the mist on/off cycle, and drainage characteristics within a nutrient mist reactor [49]. In a similar study, they developed a mathematical model for a nutrient mist reactor to cultivate *A. annua* hairy roots. Intermittent misting helps to ensure that the root surfaces are sufficiently wetted, while short cycles allow for the liquid to drain. The results of this study indicated that intermittent on/off cycles were effective for sustained operation, and to prevent water-logging of the bed, it was necessary that the mist has a short 'on' cycle. They determined that the mist could be switched off for 90 % of the run time in an optimized system [113].

5.2 Computation Fluid Dynamics (CFD)

Models of growth and mass transfer of plant tissue are necessary for bioreactor design and scale-up. However, these models ignore the hydrodynamic environment which also influences the growth and metabolism of hairy roots cultured in bioreactors. Monitoring the hydrodynamics helps to evaluate the fluid characteristics (such as fluid flow, shear stress, and turbulent intensity) that are key parameters in bioreactor design and scale-up. Computational fluid dynamics (CFD) offers a novel approach for investigating hydrodynamic behavior and designing reactor structures. CFD has been used in the simulation of hydrodynamic turbulent shear stress in a plant cell culture bioreactor [114], analysis of hydrodynamics for the design of trickle-bed reactors [115], the optimization of the inner structure in a flat algal photobioreactor [116, 117], design of the sparger in a bubble column bioreactor [118, 119], design of the impeller in a stirred tank reactor [120], and for the optimization of the height to diameter ratio in bubble column reactors [118]. The fluidization behaviors induced by ultrasound were also simulated with the help of CFD [107, 108].

5.3 CFD for Hairy Roots

Recently, Liu et al. [121] used CFD to investigate the hydrodynamics and mass transfer in an internal-loop airlift bioreactor containing *Echiniacea purpurea* hairy root cultures [122]. In this study, the CFD model was based on a porous media model (representing the hairy roots) and a discrete population balance model (representing bubble number and diameter). A two-dimensional axisymmetric model was used to describe the bioreactor due to its axial symmetric structure. The porosity of the hairy root clump was measured and calculated as follows:

$$\gamma = \frac{V - m/\rho_{root}}{V} \tag{1}$$

where V was the volume of the hairy root clump, m was the mass of the hairy root clump and ρ_{root} was density of the hairy roots of 1 g/ml.

The resistance coefficients were extrapolated after determining the pressure drop through the hairy root clumps by packing the hairy roots into a 2.54 cm × 40 cm glass tube and measuring the water flow rate under defined pressure drops. These values were used to determine the resistance coefficients D and C_2 in Eq. 2.

$$\Delta p = C_2 \frac{1}{2} \rho |\vec{u}|^2 \Delta n + D\mu |\vec{u}| \Delta n \tag{2}$$

where Δn was the porous media thickness (the glass tube height, m), ρ was the density of the fluid, and μ was the water flow rate under the pressure drop change Δp. In a test using *Echiniacea purpurea* hairy roots, parameters including liquid and gas velocity, gas holdup, mass transfer rate, and distribution of oxygen concentration in the airlift bioreactor were simulated using the CFD model. A Euler–Euler multiphase model was used to simulate the hairy root culture process. There are two types of mass transfer in the bioreactor; interphase mass transfer, in which oxygen is transferred from gas phase into liquid phase, and intraphase diffusion, the diffusion of the dissolved oxygen within the liquid phase.

Testing the parameters demonstrated that the porosity of the hairy roots changed during the culture period, from 1 to 0.7, which thereby changed the water flow resistance coefficients. As the fluid flow resistance increased, this influenced the nutrient supply for the roots. As the dissolved oxygen concentration increased, the oxygen consumption rate increased. The modeling results indicated that the dominant factors in the oxygen mass transfer in the hairy root culture were the liquid flow and turbulence. It also predicted that the concentration of dissolved oxygen would increase from the bottom to the top of the hairy root clump, and this was verified experimentally. The use of CFD provides an elegant method to guide future bioreactor design and scale-up as well as process intensification for large-scale bioreactor culture of hairy roots.

Overall, the experimental data validated the model, and provided a system that could be used to model the dissolved oxygen concentration distribution, which is difficult to determine during the process of cultivating hairy roots in a bioreactor. This model provided a simple and efficient method to predict the dissolved oxygen concentration and consumption rate distribution, and allows researchers to predict and optimize the bioreactor design and scale-up, as well as develop process intensification strategies, without having to perform numerous time-consuming bioreactor experiments.

This same model was subsequently used to simulate hydrodynamics and oxygen mass transfer in an ultrasound-intensified *Echinacea purpurea* hairy root culture [121]. The results of the simulated model, and the experimental data, demonstrated that the ultrasound intensified oxygen mass transfer in the hairy root clump, thereby stimulating growth and cichoric acid biosynthesis.

6 Conclusions

Hairy roots are an efficient system for the production of secondary metabolites, and as the global demand for these economically important compounds rises, there will a heightened interest in their production at commercial-scale. In addition to secondary metabolites, hairy roots are likely to be increasingly utilized for the production of recombinant proteins, especially for medicinal purposes. A method recently developed utilizing the combination of a chimeric super promoter, a translation enhancer sequence, and signal peptide for secretion of the recombinant protein into the media, resulted in transgenic tobacco hairy roots producing reporter protein at levels that were ten-fold higher than produced in tobacco leaves [123]. Hairy roots are also being used in phytoremediation, such as the use of *Tagetes patula* hairy roots for the decolorization of dyes [124], transgenic tobacco hairy roots for the overexpression of peroxidases for phenol removal [125], and *Brassica napus* hairy roots for the removal of 2,4-dichlorophenol from aqueous solutions [126]. Continued research in the scaled-up production of hairy roots in bioreactors will be necessary. Although methods have improved, the scale-up from shake flasks to bioreactor systems still faces many challenges; providing hairy roots with optimal levels of nutrients, oxygen, adjusting operating parameters as the culture matures, and optimizing elicitor type and timing are still being researched. In addition, in contrast to other types of culture systems, hairy roots are often morphologically varied from species to species, or even within a species, therefore optimization may be a very individual process for a specific commercial product. The continued development and improvement of low-cost and low-maintenance systems will enhance commercialization, as well as develop methods that streamline the harvesting of hairy roots from the bioreactor system. The development over the last decade, and especially in the last few years, of computational methods for the improvement of bioreactor design, such as mathematical modelling and computational fluid dynamics will be invaluable in this endeavour.

References

1. Gurib-Fakim A (2006) Medicinal plants: Traditions of yesterday and drugs of tomorrow. Mol Aspects Med 27(1):1–93
2. Bafi-Yeboa NFA, Arnason JT, Baker J, Smith ML (2005) Antifungal constituents of Northern prickly ash *Zanthoxylum americanum* Mill. Phytomedicine 12(5):370–377
3. Yoon SB, Lee YJ, Park SK, Kim HC, Bae H, Kim HM, Ko SG, Choi HY, Oh MS, Park W (2009) Anti-inflammatory effects of *Scutellaria baicalensis* water extract on LPS-activated RAW 264.7 macrophages. J Ethnopharmacol 125(2):286–290
4. Thuong PT, Pokharel YR, Lee MY, Kim SK, Bae K, Su ND, Oh WK, Kang KW (2009) Dual Anti-oxidative Effects of Fraxetin Isolated from Fraxinus rhinchophylla. Biol Pharm Bull 32(9):1527–1532
5. Yeo SK, Ooi LG, Lim TJ, Liong MT (2009) Antihypertensive properties of plant-based prebiotics. Int J Mol Sci 10(8):3517–3530
6. Efferth T, Li PCH, Konkimalla VSB, Kaina B (2007) From traditional Chinese medicine to rational cancer therapy. Trends Mol Med 13(8):353–361
7. Graham JG, Quinn ML, Fabricant DS, Farnsworth NR (2000) Plants used against cancer—an extension of the work of Jonathan Hartwell. J Ethnopharmacol 73(3):347–377
8. Bedoya LM, Sanchez-Palomino S, Abad MJ, Bermejo P, Alcami J (2001) Anti-HIV activity of medicinal plant extracts. J Ethnopharmacol 77(1):113–116
9. Liu CZ, Towler MJ, Medrano G, Cramer CL, Weathers PJ (2009) Production of mouse interleukin-12 is greater in tobacco hairy roots grown in a mist reactor than in an airlift reactor. Biotechnol Bioeng 102(4):1074–1086
10. Woodard SL, Wilken LR, Barros GOF, White SG, Nikolov ZL (2009) Evaluation of monoclonal antibody and phenolic extraction from transgenic Lemna for purification process development. Biotechnol Bioeng 104(3):562–571
11. Verpoorte R, Contin A, Memelink J (2002) Biotechnology for the production of plant secondary metabolites. Phytochem Rev 1(1):13–25
12. Chilton MD, Tepfer DA, Petit A, David C, Cassedelbart F, Tempe J (1982) *Agrobacterium rhizogenes* inserts T-DNA into the genomes of the host plant-root cells. Nature 295(5848):432–434
13. Mishra BN, Ranjan R (2008) Growth of hairy-root cultures in various bioreactors for the production of secondary metabolites. Biotechnol Appl Biochem 49:1–10
14. Hu ZB, Du M (2006) Hairy root and its application in plant genetic engineering. J Integr Plant Biol 48:121–127
15. Hamill JD, Robins RJ, Parr AJ, Evans DM, Furze JM, Rhodes MJC (1990) Over-expressing a yeast ornithine decarboxylase gene in transgenic Roots of *Nicotiana rustica* Can Lead to Enhanced Nicotine Accumulation. Plant Mol Biol 15(1):27–38
16. Hashimoto T, Yun DJ, Yamada Y (1993) Production of tropane alkaloids in genetically engineered root cultures. Phytochemistry 32(3):713–718

17. Moyano E, Jouhikainen K, Tammela P, Palazon J, Cusido RM, Pinol MT, Teeri TH, Oksman-Caldentey KM (2003) Effect of pmt gene overexpression on tropane alkaloid production in transformed root cultures of *Datura metel* and *Hyoscyamus muticus*. J Exp Bot 54(381):203–211
18. Zhang L, Ding RX, Chai YR, Bonfill M, Moyano E, Oksman-Caldentey KM, Xu TF, Pi Y, Wang ZN, Zhang HM, Kai GY, Liao ZH, Sun XF, Tang KX (2004) Engineering tropane biosynthetic pathway in *Hyoscyamus niger* hairy root cultures, In: Proceedings of the national academy of sciences of the United States of America 101(17):6786–6791
19. Jouhikainen K, Lindgren L, Jokelainen T, Hiltunen R, Teeri TH, Oksman-Caldentey KM (1999) Enhancement of scopolamine production in *Hyoscyamus muticus* L. hairy root cultures by genetic engineering. Planta 208(4):545–551
20. Li FX, Jin ZP, Zhao DX, Cheng LQ, Fu CX, Ma FS (2006) Overexpression of the *Saussurea medusa* chalcone isomerase gene in *Saussurea involucrata* hairy root cultures enhances their biosynthesis of apigenin. Phytochemistry 67(6):553–560
21. Huang TK, McDonald KA (2012) Bioreactor systems for in vitro production of foreign proteins using plant cell cultures. Biotechnol Adv 30(2):398–409
22. Sevon N, Drager B, Hiltunen R, OksmanCaldentey KM (1997) Characterization of transgenic plants derived from hairy roots of *Hyoscyamus muticus*. Plant Cell Rep 16(9):605–611
23. Shanks JV, Morgan J (1999) Plant hairy root culture. Curr Opin Biotechnol 10(2):151–155
24. Kim Y, Wyslouzil BE, Weathers PJ (2002a) Invited review: secondary metabolism of hairy root cultures in bioreactors. In Vitro Cell Dev Biol Plant 38(1):1–10
25. Eibl R, Eibl D (2008) Design of bioreactors suitable for plant cell and tissue cultures. Phytochem Rev 7:593–598
26. Paek KY, Chakrabarty D, Hahn EJ (2005) Application of bioreactor systems for large scale production of horticultural and medicinal plants. Plant Cell, Tissue Organ Cult 81(3):287–300
27. Curtis WR (2000) Bioreactor growth of hairy roots. In: Spier R (ed) Encyclopedia of Cell Technology. John Wiley and Sons, New York, pp 827–841
28. Jung G, Tepfer D (1987) Use of Genetic-Transformation by the Ri T-DNA of *Agrobacterium rhizogenes* to stimulate biomass and tropane alkaloid production in *Atropa belladonna* and Calystegia sepium roots grown-invitro. Plant Sci 50(2):145–151
29. Wilson PDG, Hilton MG, Robins RJ, Rhodes MJC (1987) Fermentation studies of transformed root cultures. In: international conference on bioreactors and biotransformations, Elsevier, London, pp 38–51
30. Doran PM (1999) Design of mixing systems for plant cell suspensions in stirred reactors. Biotechnol Prog 15(3):319–335
31. Kondo O, Honda H, Taya M, Kobayashi T (1989) Comparison of growth-properties of carrot hairy root in various bioreactors. Appl Microbiol Biotechnol 32(3):291–294
32. Uozumi N, Kohketsu K, Kobayashi T (1993) Growth and kinetic-parameters of Ajuga hairy root in fed-batch culture on monosaccharide medium. J Chem Technol Biotechnol 57(2):155–161
33. Davioud E, Kan C, Hamon J, Tempe J, Husson HP (1989) Production of indole alkaloids by In vitro root cultures from *Catharanthus trichophyllus*. Phytochemistry 28(10):2675–2680
34. Jeong GT, Park DH, Hwang B, Woo JC (2003) Comparison of growth characteristics of *Panax ginseng* hairy roots in various bioreactors. Appl Biochem Biotechnol 105:493–503
35. Srivastava S, Srivastava AK (2012b) In vitro Azadirachtin production by hairy root cultivation of *Azadirachta indica* in nutrient mist bioreactor. Appl Biochem Biotechnol 166(2):365–378
36. Srivastava S, Srivastava AK (2012a) Azadirachtin production by hairy root cultivation of *Azadirachta indica* in a modified stirred tank reactor. Bioprocess Biosystems Engineering 35(9):1549–1553

37. Choi Y, Kim Y, Paek K (2006) Types and designs of bioreactors for hairy culture. In: Duttagupta S, Ibaraki Y (eds) Plant tissue culture engineering. Springer, Dordrecht, pp 161–172
38. Kwok KH, Doran PM (1995) Kinetic and stoichiometric analysis of hairy roots in a segmented bubble-column reactor. Biotechnol Prog 11(4):429–435
39. Kanokwaree K, Doran PM (1998) Application of membrane tubing aeration and perfluorocarbon to improve oxygen delivery to hairy root cultures. Biotechnol Prog 14(3):479–486
40. Min JY, Jung HY, Kang SM, Kim YD, Kang YM, Park DJ, Prasad DT, Choi MS (2007) Production of tropane alkaloids by small-scale bubble column bioreactor cultures of *Scopolia parviflora* adventitious roots. Bioresour Technol 98(9):1748–1753
41. Yoshikawa T, Furuya T (1987) Saponin production by cultures of *Panax ginseng* transformed with *Agrobacterium rhizogenes*. Plant Cell Reps 6(6):449–453
42. Taya M, Yoyama A, Kondo O, Kobayashi T, Matsui C (1989) Growth-characteristics of plant hairy roots and their cultures in bioreactors. J Chem Eng Japan 22(1):84–89
43. Rodriguez-Mendiola MA, Stafford A, Cresswell R, Ariascastro C (1991) Bioreactors for growth of plant-roots. Enzym Microb Technol 13(9):697–702
44. Sauerwein M, Yamazaki T, Shimomura K (1991) Hernandulcin in hairy root cultures of *Lippia dulcis*. Plant Cell Rep 9(10):579–581
45. Shimomura K, Sudo H, Saga H, Kamada H (1991) Shikonin production and secretion by hairy root cultures of *Lithospermum erythrorhizon*. Plant Cell Rep 10(6–7):282–285
46. Sudo H, Yamakawa T, Yamazaki M, Aimi N, Saito K (2002) Bioreactor production of camptothecin by hairy root cultures of *Ophiorrhiza pumila*. Biotechnol Lett 24(5):359–363
47. Abbasi BH, Liu R, Saxena PK, Liu CZ (2009) Cichoric acid production from hairy root cultures of *Echinacea purpurea* grown in a modified airlift bioreactor. J Chem Technol Biotechnol 84(11):1697–1701
48. Kintzios S, Makri O, Pistola E, Matakiadis T, Shi HP, Economou A (2004) Scale-up production of puerarin from hairy roots of *Pueraria phaseoloides* in an airlift bioreactor. Biotechnol Lett 26(13):1057–1059
49. Katuri SR, Ranjan R, Khanna R (2011) Mathematical modeling of mist bioreactor for the growth of hairy roots In: National workshop-cum-conference on recent trends in Mathematics and Computing
50. Ramakrishnan D, Salim J, Curtis WR (1994) Inoculation and tissue distribution in pilot-scale plant-root culture bioreactors. Biotechnol Tech 8(9):639–644
51. Wyslouzil BE, Whipple M, Chatterjee C, Walcerz DB, Weathers PJ, Hart DP (1997) Mist deposition onto hairy root cultures: Aerosol modeling and experiments. Biotechnol Prog 13(2):185–194
52. Towler MJ, Kim Y, Wysiouzil BE, Correll MJ, Weathers PJ (2006) Design, development, and applications of mist bioreactors for micropropagation and hairy root culture. In: Gupta SD, Ibaraki Y (eds) Plant tissue culture engineering. Springer, Netherlands, pp 119–134
53. Sivakumar G, Liu CZ, Towler M, Weathers PJ (2010b) Biomass production of hairy roots of *Artemisia annua* and *Arachis hypogaea* in a scaled-up mist bioreactor. Biotechnol Bioeng 107(5):802–813
54. Kochan E, Krolicka A, Chmiel A (2012) Growth and Ginsenoside production in *Panax quinquefolium* hairy roots cultivated in flasks and nutrient sprinkle bioreactor. Acta Physiologiae Plantarum 34(4):1513–1518
55. Grzegorczyk I, Wysokinska H (2010) Antioxidant compounds in *Salvia officinalis* L. shoot and hairy root cultures in the nutrient sprinkle bioreactor. Acta Societatis Botanicorum Poloniae 79(1):7–10
56. Kuzma L, Bruchajzer E, Wysokinska H (2009) Methyl jasmonate effect on diterpenoid accumulation in *Salvia sclarea* hairy root culture in shake flasks and sprinkle bioreactor. Enzym Microb Technol 44(6–7):406–410
57. Williams GRC, Doran PM (1999) Investigation of liquid-solid hydrodynamic boundary layers and oxygen requirements in hairy root cultures. Biotechnol Bioeng 64(6):729–740

58. Ramakrishnan D, Curtis WR (2004) Trickle-bed root culture bioreactor design and scale-up: Growth, fluid-dynamics, and oxygen mass transfer. Biotechnol Bioeng 88(2):248–260
59. Kim Y, Wyslouzil BE, Weathers PJ (2001) A comparative study of mist and bubble column reactors in the in vitro production of Artemisinin. Plant Cell Rep 20(5):451–455
60. Kim YJ, Weathers PJ, Wyslouzil BE (2002c) Growth of *Artemisia annua* hairy roots in liquid- and gas-phase reactors. Biotechnol Bioeng 80(4):454–464
61. Suresh B, Bais HP, Raghavarao KSMS, Ravishankar GA, Ghildyal NP (2005) Comparative evaluation of bioreactor design using *Tagetes patula* L. hairy roots as a model system. Process Biochem 40(5):1509–1515
62. Wilson PDG (1997) The pilot-scale cultivation of transformed roots. In: Doran PM (ed) Hairy roots: culture and applications. Harwood Academic, Amsterdam, pp 190–197
63. Eibl R, Eibl D (2006) Design and use of the wave bioreactor for plant cell culture. In: Gupta SD, Ibaraki Y (eds) Plant tissue culture engineering. Springer, Netherlands, pp 203–227
64. Asplund PT, Curtis WR (2001) Intrinsic oxygen use kinetics of transformed plant root culture. Biotechnol Prog 17(3):481–489
65. Kim Y, Wyslouzil BE, Weathers PJ (2002b) Invited review: secondary metabolism of hairy root cultures in bioreactors. In vitro Cell Dev Biol Plant 38(1):1–10
66. Bordonaro JL, Curtis WR (2000) Inhibitory role of root hairs on transport within root culture bioreactors. Biotechnol Bioeng 70(2):176–186
67. Shiao TL, Doran PM (2000) Root hairiness: effect on fluid flow and oxygen transfer in hairy root cultures. J Biotechnol 83(3):199–210
68. Dhaouadi H, Poncin S, Hornut JM, Midoux N (2008) Gas-liquid mass transfer in bubble column reactor: analytical solution and experimental confirmation. Chem Eng Process 47(4):548–556
69. Ducos JP, Terrier B, Courtois D (2009) Disposable bioreactors for plant micropropagation and mass plant cell culture. Dispos Bioreactors 115:89–115
70. Carvalho EB, Curtis WR (1998) Characterization of fluid-flow resistance in root cultures with a convective flow tubular bioreactor. Biotechnol Bioeng 60(3):375–384
71. Hitaka Y, Kino-Oka M, Taya M, Tone S (1997) Effect of liquid flow on culture of red beet hairy roots in single column reactor. J Chem Eng Jpn 30(6):1070–1075
72. Kino-Oka R, Hitaka Y, Taya M, Tone S (1999) High-density culture of red beet hairy roots by considering medium flow condition in a bioreactor. Chem Eng Sci 54(15–16):3179–3186
73. Shiao TI, Ellis MH, Dolferus R, Dennis ES, Doran PM (2002) Overexpression of alcohol dehydrogenase or pyruvate decarboxylase improves growth of hairy roots at reduced oxygen concentrations. Biotechnol Bioeng 77(4):455–461
74. Dilorio AA, Cheetham RD, Weathers PJ (1992) Carbon-dioxide improves the growth of hairy roots cultured on solid medium and in nutrient mists. Appl Microbiol Biotechnol 37(4):463–467
75. Wyslouzil BE, Waterbury RG, Weathers PJ (2000) The growth of single roots of *Artemisia annua* in nutrient mist reactors. Biotechnol Bioeng 70(2):143–150
76. Sung LS, Huang SY (2000) Headspace ethylene accumulation on *Stizolobium hassjoo* hairy root culture producing L-3,4-dihydroxyphenylalanine. Biotechnology Letters 22(10):875–878
77. Abbasi BH, Tian CL, Murch SJ, Saxena PK, Liu CZ (2007) Light-enhanced caffeic acid derivatives biosynthesis in hairy root cultures of *Echinacea purpurea*. Plant Cell Rep 26(8):1367–1372
78. Wang YC, Zhang HX, Zhao B, Yuan XF (2001) Improved growth of *Artemisia annua* L hairy roots and artemisinin production under red light conditions. Biotechnol Lett 23(23):1971–1973
79. Zhong JJ, Seki T, Kinoshita S, Yoshida T (1991) Effect of light irradiation on Anthocyanin production by suspended culture of *Perilla frutescens*. Biotechnol Bioeng 38(6):653–658
80. Taya M, Sato H, Kinooka M, Tone S (1994) Characterization of Pak-bung green hairy roots cultivated under light irradiation. J Ferment Bioeng 78(1):42–48

81. Bhadra R, Morgan JA, Shanks JV (1998) Transient studies of light-adapted cultures of hairy roots of *Catharanthus roseus*: Growth and indole alkaloid accumulation. Biotechnol Bioeng 60(6):670–678
82. Jacob A, Malpathak N (2004) Green hairy root cultures of *Solanum khasianum* clarke—a new route to in vitro Solasodine production. Curr Sci 87(10):1442–1447
83. Flores H, Medina-Bolivar H (1993) Root cultures and plant natural products: unearthing the hidden half of plant metabolism. In: Plant tissue culture and biotechnology. Balaban Publisher, UK, pp 59–74
84. Liu CZ, Guo C, Wang YC, Ouyang F (2002) Effect of light irradiation on hairy root growth and Artemisinin biosynthesis of *Artemisia annua* L. Process Biochem 38(4):581–585
85. Yu KW, Murthy HN, Hahn EJ, Paek KY (2005) Ginsenoside production by hairy root cultures of *Panax ginseng*: influence of temperature and light quality. Biochem Eng J 23(1):53–56
86. Yu SX, Kwok KH, Doran PM (1996) Effect of sucrose, exogenous product concentration, and other culture conditions on growth and steroidal alkaloid production by *Solanum aviculare* hairy roots. Enzym Microb Technol 18(4):238–243
87. Hilton MG, Rhodes MJC (1990) Growth and Hyoscyamine production of hairy root cultures of *Datura stramonium* in a modified stirred tank reactor. Appl Microbiol Biotechnol 33(2):132–138
88. Sivakumar G, Yu KW, Hahn EJ, Paek KY (2005) Optimization of organic nutrients for ginseng hairy roots production in large-scale bioreactors. Curr Sci 89(4):641–649
89. Wilhelmson A, Hakkinen ST, Kallio PT, Oksman-Caldentey KM, Nuutila AM (2006) Heterologous expression of Vitreoscilla hemoglobin (VHb) and cultivation conditions affect the alkaloid profile of *Hyoscyamus muticus* hairy roots. Biotechnol Prog 22(2):350–358
90. Weathers PJ, Hemmavanh DD, Walcerz DB, Cheetham RD, Smith TC (1997) Interactive effects of nitrate and phosphate salts, sucrose, and inoculum culture age on growth and sesquiterpene production in *Artemisia annua* hairy root cultures. In vitro Cell Dev Biol Plant 33(4):306–312
91. Amdoun R, Khelifi L, Khelifi-Slaoui M, Amroune S, Benyoussef EH, Thi DV, Assaf-Ducrocq C, Gontier E (2009) Influence of minerals and elicitation on *Datura stramonium* L. tropane alkaloid production: Modelization of the in vitro biochemical response. Plant Sci 177(2):81–87
92. Cui XH, Chakrabarty D, Lee EJ, Paek KY (2010) Production of adventitious roots and secondary metabolites by *Hypericum perforatum* L in a bioreactor. Bioresour Technol 101(12):4708–4716
93. Sreedhar RV, Roohie K, Maya P, Venkatachalam L, Bhagyalakshmi N (2009) Biotic elicitors enhance flavour compounds during accelerated curing of vanilla beans. Food Chem 112(2):461–468
94. Vasconsuelo A, Boland R (2007) Molecular aspects of the early stages of elicitation of secondary metabolites in plants. Plant Sci 172(5):861–875
95. Georgiev MI, Pavlov AI, Bley T (2007) Hairy root type plant in vitro systems as sources of bioactive substances. Appl Microbiol Biotechnol 74(6):1175–1185
96. Pitta-Alvarez SI, Spollansky TC, Giulietti AM (2000) The influence of different biotic and abiotic elicitors on the production and profile of tropane alkaloids in hairy root cultures of *Brugmansia candida*. Enzym Microb Technol 26(2–4):252–258
97. Satdive RK, Fulzele DP, Eapen S (2007) Enhanced production of Azadirachtin by hairy root cultures of *Azadirachta indica* A Juss by elicitation and media optimization. J Biotechnol 128(2):281–289
98. Rahimi S, Hasanloo T, Najafi F, Khavari-Nejad RA (2011) Methyl jasmonate influence on Silymarin production and plant stress responses in *Silybum marianum* hairy root cultures in a bioreactor. Nat Prod Res 26(18):1662–1667
99. Putalun W, Luealon W, De-Eknamkul W, Tanaka H, Shoyama Y (2007) Improvement of Artemisinin production by Chitosan in hairy root cultures of *Artemisia annua* L. Biotechnol Lett 29(7):1143–1146

100. Savitha BC, Thimmaraju R, Bhagyalakshmi N, Ravishankar GA (2006) Different biotic and abiotic elicitors influence Betalain production in hairy root cultures of *Beta vulgaris* in shake-flask and bioreactor. Process Biochem 41(1):50–60
101. Bhagwath SG, Hjortso MA (2000) Statistical analysis of elicitation strategies for Thiarubrine A production in hairy root cultures of *Ambrosia artemisiifolia*. J Biotechnol 80(2):159–167
102. Lin LD, Wu JY (2002) Enhancement of shikonin production in single- and two-phase suspension cultures of *Lithospermum erythrorhizon* cells using low-energy ultrasound. Biotechnol Bioeng 78(1):81–88
103. Lin LD, Wu JY, Ho KP, Qi SY (2001) Ultrasound-induced physiological effects and secondary metabolite (Saponin) production in *Panax ginseng* cell cultures. Ultrasound Med Biol 27(8):1147–1152
104. Wu J, Lin L (2003) Enhancement of Taxol production and release in *Taxus chinensis* cell cultures by ultrasound, methyl jasmonate and in situ solvent extraction. Appl Microbiol Biotechnol 62(2–3):151–155
105. Chisti Y (2003) Sonobioreactors: using ultrasound for enhanced microbial productivity. Trends Biotechnol 21(2):89–93
106. Kumar A, Kumaresan T, Pandit AB, Joshi JB (2006) Characterization of flow phenomena induced by ultrasonic horn. Chem Eng Sci 61(22):7410–7420
107. Cao CQ, Dong SQ, Zhao YA, Guo QJ (2010) Experimental and numerical research for fluidization behaviors in a gas-solid acoustic fluidized bed. AICHE J 56(7):1726–1736
108. Laborde JL, Hita A, Caltagirone JP, Gerard A (2000) Fluid dynamics phenomena induced by power ultrasounds. Ultrasonics 38(1–8):297–300
109. Kim S, Hopper E, Hjortso M (1995) Hairy root-growth models—effect of different branching patterns. Biotechnol Prog 11(2):178–186
110. Lenk F, Vogel M, Bley T, Steingroewer J (2012) Automatic image recognition to determine morphological development and secondary metabolite accumulation in hairy root networks. Eng Life Sci 12(5):1–7
111. Bastian P, Chavarria-Krauser A, Engwer C, Jager W, Marnach S, Ptashnyk M (2008) Modelling in vitro growth of dense root networks. J Theor Biol 254(1):99–109
112. Prince CL, Bringi V, Shuler ML (1991) Convective mass-transfer in large porous biocatalysts—plant organ-cultures. Biotechnol Prog 7(2):195–199
113. Ranjan R, Khanna R, Mishra BN (2011) Sustained operation of nutrient mist reactor to grow hairy roots. Asia-Pacific J Chem Eng 6(1):23–28
114. Sowana DD, Williams DRG, Dunlop EH, Dally BB, O'Neill BK, Fletcher DF (2001) Turbulent shear stress effects on plant cell suspension cultures. Chem Eng Res Des 79(A8):867–875
115. Atta A, Roy S, Nigam KDP (2010) A two-phase Eulerian approach using relative permeability concept for modeling of hydrodynamics in trickle-bed reactors at elevated pressure. Chem Eng Res Des 88(3):369–378
116. Merchuk J, Garcia-Camacho F, Molina-Grima E (2007) Photobioreactor design and fluid dynamics. Chem Biochem Eng Quarterly 21(4):345–355
117. Yu G, Li YC, Shen GM, Wang WL, Lin C, Wu HX, Chen ZS (2009) A novel method using CFD to optimize the inner structure parameters of flat photobioreactors. J Appl Phycol 21(6):719–727
118. Dhotre MT, Ekambara K, Joshi JB (2004) CFD simulation of sparger design and height to diameter ratio on gas hold-up profiles in bubble column reactors. Exp Thermal Fluid Sci 28(5):407–421
119. Zhong C, Yuan YJ (2009) Responses of *Taxus cuspidata* to hydrodynamics in bubble column bioreactors with different sparging nozzle sizes. Biochem Eng J 45(2):100–106
120. Ding J, Wang X, Zhou XF, Ren NQ, Guo WQ (2010) CFD optimization of continuous stirred-tank (CSTR) reactor for biohydrogen production. Bioresour Technol 101(18):7005–7013

121. Liu R, Sun W, Liu CZ (2011a) Computational fluid dynamics modeling of mass-transfer behavior in a bioreactor for hairy root culture. II. Analysis of ultrasound-intensified process. Biotechnol Prog 27(6):1672–1679
122. Liu R, Sun W, Liu CZ (2011b) Computational fluid dynamics modeling of mass transfer behavior in a bioreactor for hairy root culture. I. Model development and experimental validation. Biotechnol Prog 27(6):1661–1671
123. Ñopo L, Woffenden BJ, Reed DG, Buswell S, Zhang C, Medina-Bolivar F (2012) Super-promoter: TEV, a powerful gene expression system for tobacco hairy Roots. In: Lorence Argelia (ed) Recombinant gene expression: reviews and protocols, methods in molecular biology. Springer, Science + Business Media, pp 501–526
124. Patil P, Desai N, Govindwar S, Jadhav JP, Bapat V (2009) Degradation analysis of reactive red 198 by hairy roots of *Tagetes patula* L. (Marigold). Planta 230(4):725–735
125. Alderete LGS, Talano MA, Ibáñez SG, Purro S, Agostini E, Milrad SR, Medina MI (2009) Establishment of transgenic tobacco hairy roots expressing basic peroxidises and its application for phenol removal. J Biotechnol 139(4):273–279
126. Angelini VA, Orejas J, Medina MI, Agostini E (2011) Scale up of 2,4-dichlorophenol removal from aqueous solutions using *Brassica napus* hairy roots. J Hazard Mater 185(1):269–274

Adv Biochem Eng Biotechnol (2013) 134: 115–134
DOI: 10.1007/10_2013_184

Published Online: 7 May 2013

Hairy Roots as a Vaccine Production and Delivery System

Marina Skarjinskaia, Karen Ruby, Adriana Araujo, Karina Taylor, Vengadesan Gopalasamy-Raju, Konstantin Musiychuk, Jessica A. Chichester, Gene A. Palmer, Patricia de la Rosa, Vadim Mett, Natalia Ugulava, Stephen J. Streatfield and Vidadi Yusibov

Abstract Prevention of infectious diseases by vaccination is often limited because of the lack of safe, effective, and accessible vaccines. Traditional vaccines are expensive and require special conditions for storage, distribution, and administration. Plants have potential for large-scale production of a variety of inexpensive and highly effective recombinant proteins for biomedical and pharmaceutical applications, including subunit vaccines. There are several approaches for the production of vaccine antigens in plants, including transient expression systems based on Agrobacterium delivery of binary vectors or plant viral vectors, stable transgenic plants, and plant cell or tissue cultures. Axenic plant cultures maintained under defined physical and chemical conditions appear to be an attractive production platform when target proteins need to be synthesized in a fully controlled environment. Hairy root cultures meet the criteria for such a system. Hairy root cultures, generated from edible plants and producing target antigens, provide a potential approach for the development of vaccines for oral delivery. With this approach, there are no protein extraction and purification costs and the active biomolecule is protected by the plant cell wall during passage through the upper gastrointestinal tract. This allows for gradual release of antigen at mucosal surfaces in the gut. Lyophilized hairy root cultures expressing vaccine antigens can be stored at ambient temperature for extended periods of time, which should facilitate storage and distribution, ultimately allowing for large populations to be vaccinated.

Keywords Hairy roots · Oral immunization · Recombinant vaccine · Transgenic plant · Viral vector

---

M. Skarjinskaia · K. Ruby · A. Araujo · K. Taylor · V. Gopalasamy-Raju · K. Musiychuk · J. A. Chichester · G. A. Palmer · P. de la Rosa · V. Mett · N. Ugulava · S. J. Streatfield · V. Yusibov (✉)
Fraunhofer USA Center for Molecular Biotechnology, 9 Innovation Way, Suite 200, Newark DE, 19711, USA
e-mail: vyusibov@fraunhofer-cmb.org

Contents

1 Introduction

In the last two decades, a number of groups from both academia and industry have demonstrated the potential of plants as a promising platform for producing recombinant proteins, including those for human health applications. Bacterial, yeast, insect, and/or mammalian expression systems are effective for certain targets, but each has particular limitations associated with the costs of manufacturing, the system's ability to produce soluble and complex targets and perform post-translational modifications (PTMs), and product safety [1, 2]. In contrast, plant-based expression systems are highly scalable, and the target product is safe due to the host being free of animal pathogens or microbial toxins [3]. Moreover, PTMs in plants are broadly similar to those of mammalian systems, with minor differences in glycan residue decoration not appearing to affect specific immunogenicity of the target [2, 4, 5].

With the success in the development of plant-based expression systems, vaccination with plant tissue-based mucosal vaccines has been considered an attractive alternative to parenteral administration of purified recombinant antigens. Moreover, although parenterally administered vaccines generally elicit strong serum IgG antibodies, they do not typically induce robust mucosal secretory immunoglobulin type A (sIgA) production [6, 7] and therefore may be unable to protect against many pathogens that colonize mucosal surfaces of the gastrointestinal (GI), respiratory, or genital tracts. By contrast, oral administration represents an effective route for stimulating local mucosal immune responses in the oral cavity and the GI tract, potentially leading to robust generalized mucosal immune responses [8]. For oral administration of a plant-based edible vaccine, minimal processing of plant material is required to ensure uniform dosing and stable product, and the antigen is afforded natural bioencapsulation and consequent stability in the GI tract, a typically hostile environment for delivered antigen [9]. Although there are examples of edible vaccines inducing immune responses without adjuvant, mucosal immunity can be enhanced by using appropriate adjuvants [10–13], including noncatalytic subunits of *Escherichia coli* heat-labile toxin [14] or *Vibrio cholera* [15], as well as adjuvants of plant origin, such as lectins [16, 17].

On the other hand, edible plant-based vaccines have some disadvantages, including low levels of target antigen expression, necessitating consumption of large quantities of product to stimulate immune responses [18], and limited regulatory and human experience. Therefore, to successfully launch an orally administered plant tissue-based vaccine, issues related to economics, safety, and potency should be addressed. An expression system must be able to produce sufficient quantities of target antigen with the desired consistency, stability, safety, and potency, including storage stability at elevated temperatures as well as stability against enzymatic digestion in the GI tract following vaccine administration [9]. Transgenic edible plants have been the focus of most reported studies [6, 19–21].

As demonstrated in a number of studies, antigens from various pathogens, including enterotoxigenic *E. coli* (ETEC; [22]), measles virus [23], *Plasmodium falciparum* [24], *Vibrio cholera* [25], foot-and-mouth disease virus [26], hepatitis B virus [27, 28], and rabies virus [29], have been expressed in plants, tested in animal models, and shown to be immunogenic when administered orally. Furthermore, plant-derived edible vaccines against ETEC [30], hepatitis B virus [29, 31], and rabies virus [32] have been evaluated in clinical trials with demonstrated safety and immunogenicity in humans. Despite the successful production of recombinant proteins in transgenic plants, biosafety and regulatory issues are major concerns [33, 34]. Therefore, contained expression systems such as plant cell cultures or cultures of differentiated organs, such as hairy roots, are being considered as alternatives to transgenic whole plant-based systems.

2 Generation of Hairy Root Cultures for Producing Heterologous Proteins

Hairy roots are neoplastic tissues that develop as a result of expression of *rol* and *aux* genes of *Agrobacterium rhizogenes* transformed into the host cell genome following infection with this Gram-negative soil bacterium [35]. Hairy root development is associated with the formation of extensive secondary roots that can be excised and cultivated indefinitely under sterile conditions. Hairy root cultures (HRCs) are maintained in media containing a mixture of salts and sucrose and lacking products of animal origin. In addition, HRC media are hormone-free, which eliminates the possibility of hormone-induced genotype instability through chromosome aberrations [36, 37]. HRCs have been established from numerous species of dicotyledonous plants including those with attractive phytochemical [38] or metabolic properties [39–43].

HRC-based production systems, under contained conditions, offer substantial benefits, including fast economic biomass accumulation, genetic stability, differentiated organ-specific cultures of clonal origin [44], and time efficiency. It requires 50–60 days to obtain a master culture followed by biomass increase correlating with doubling time of the HRC. For example, mean doubling time of the HRC depends on plant species and ranges from 1.2 (*Atropa belladonna*) to 7.7 days (*Solanum aviculare*; [45]).

HRCs show consistency in target gene expression over an extended period of time. The examples include full-length murine IgG1 monoclonal antibody produced in tobacco HRCs with stable expression for about 19 months [46] and β-glucuronidase (GUS) produced in *Lotus corniculatus* hairy roots with stable expression over 5 years [47]. In addition, consistent production level of green fluorescent protein cycle 3 (GFPC3) or human growth factor (hGH) was recorded for a 3-year period in *Nicotiana benthamiana* HRCs where expression of the target proteins was driven by an episomal expression vector [48]. Similar genotypic stability was observed for secondary metabolite production in HRCs. For example, Marconi et al. [49] reported on stable production of tropane alkaloids by *Brugmansia candida* HRCs for 5 years. Baiza et al. [50] observed stable production of tropane alkaloids in HRCs of *Datura stramonium* over a 6-year period.

HRCs have been demonstrated to express recombinant proteins at up to three- to fivefold higher levels compared to their "parental" transgenic plant lines, as shown for acetylcholinesterase-expressing HRCs generated from transgenic *N. benthamiana* plants [51] and hepatitis B surface antigen (HBsAg)-expressing HRCs derived from transgenic potato [52]. Secondary metabolites are also produced by HRCs at an up to twofold higher level than their "parental" plant lines [53–56].

Hairy roots are able to secrete properly folded functionally active recombinant proteins and natural products into the culture medium [46, 57]. Secretory signal sequences directing a protein of interest to the hairy root secretory pathway can be exploited in secretion-based aseptic in vitro systems where isolation of recombinant proteins from hydroponic media can be relatively cost effective [58].

The bottleneck for the application of HRCs for the production of recombinant proteins or natural products is the establishment of effective and economic processes for scale-up, despite several bioreactor designs having been developed [59].

HRCs, genetically modified, have the potential to be used as a platform for the production of target proteins [42, 60–62], including monoclonal antibodies [46, 63] and vaccine antigens [52, 64–68]. Below we review different vector systems used for the expression of recombinant pharmaceutical proteins in HRCs.

3 Approaches for Expressing Foreign Proteins in Hairy Root Cultures

3.1 Transgenic

As previously described in Sect. 2, wild-type HRCs are commonly generated from transgenic plants by infecting them with *A. rhizogenes* or by transforming plants with *A. rhizogenes* genetically modified by introducing disarmed T-DNA of a binary plasmid ([69]; Fig. 1), which is a standard tool for integration of the transgene into

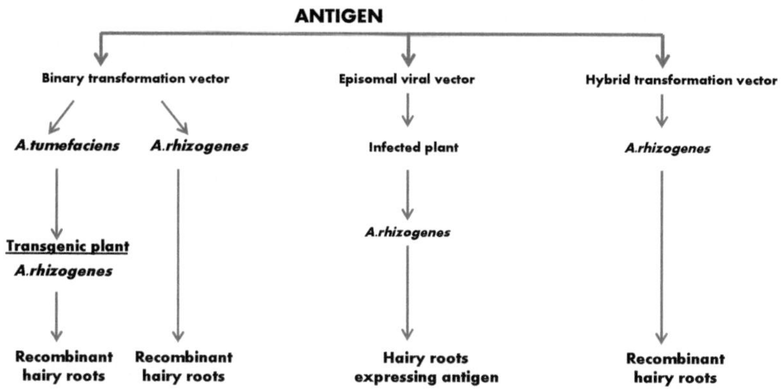

Fig. 1 Schematic diagram of hairy root culture systems for heterologous antigen expression

the plant nuclear genome [70]. One of the key limitations of transgenic hairy roots is a weak target expression. Efforts are being made to increase target accumulation in HRCs by including strong regulatory elements in expression cassettes for binary vectors [71], for example, the constitutive 35S *Cauliflower mosaic virus* (CaMV) promoter [72], the *ocs-mas2′* super-promoter with the highest activity in roots of transgenic tobacco plants [73], or strong root-specific promoters such as the promoter from the tomato *SlREO* gene [74]. Constitutive promoters are often preferred; however, due to the toxic nature of some recombinant proteins, constitutive expression can be detrimental or lethal to hairy roots [68]. By contrast, inducible promoters allow for expressing only after specific treatment with physical stimuli such as altered temperature [75], change in light spectrum or regimen [76], or wounding [77], or chemical stimuli such as metal ions [78], alcohols [79, 80, 85], steroids [81], herbicides [82], or insecticides [83, 84]. Thus, application of an inducible system allows for separating the "growth" and "production" phases for recombinant protein manufacturing in plants [85].

Inclusion of viral translational enhancers such as the *omega* 5' leader sequence of *tobacco mosaic virus* (TMV; [86]), optimization of codon usage to that preferred by the plant host ([71, 87]), and the use of subcellular targeting sequences such as the endoplasmic reticulum (ER) targeting sequence, KDEL, can boost target expression [88, 89]. Additionally, to enhance target solubility and stability, fusion with carrier molecules such as β-1,3-1,4-glucanase from a thermophilic bacterium *Clostridium thermocellum* (lichenase; [90]) has been applied.

3.2 Transient

In addition to the stable incorporation into the plant cell genome, target genes can be expressed in hairy roots using recombinant plant viral vectors. Due to the ability of plant viral RNA to self-replicate in the cytoplasm of infected plant cells and move into neighboring cells [91, 92], plant viruses can infect and continually replicate in HRCs. For example, Shadwick and Doran [93] established viral infection of hairy roots by co-incubation with a TMV-based vector, TMV-30B, carrying a fluorescent tag, GFPC3 [94]. After the recombinant virus was added to the culture medium containing hairy roots, virus accumulation was observed in growing root biomass, implying cell-to-cell movement of the virus within the root tissue [93]. However, the virus titer in the hairy root biomass declined during cultivation, most likely because of the genetic instability of the TMV-30B viral vector [94]. Furthermore, no GFP expression was detected in virus-infected root cultures, likely because of the loss of GFP-encoding sequences during passaging of the viral vector from leaves.

The feasibility of target protein production in HRCs mediated by replication of recombinant virus without nuclear integration of target sequences has also been demonstrated by Skarjinskaia et al. [48]. In this study, the TMV-30B vector encoding either hGH or GFPC3 was introduced into cells of greenhouse-grown *N. benthamiana* by mechanical inoculation. Hairy roots were then induced from leaves that were systemically infected by the virus by applying wild-type *A. rhizogenes* (Fig. 1). Among emerged roots, 70 % were found to express target protein. Accumulation of both hGH and GFP in selected clonal root lines was sustained over a 3-year period in the absence of selection and at levels comparable with those determined at the time of establishing the hairy root lines. Reports have demonstrated that virus infection did not affect the hairy root proliferation and biomass gain [48, 93].

To demonstrate the stability of the recombinant virus in HRCs, *N. benthamiana* plants were inoculated with extract from a 3-year old hairy root line expressing GFP (Skarjinskaia et al. unpublished observation). Following inoculation, viral replication and GFP expression were observed throughout the plant in locally inoculated and systemically infected leaves. Conversely, when the 30B-GFP vector was passaged in plants, GFP expression was lost after two to three serial passages (4–6 weeks), probably due to virus recombination [95].

Recombinant proteins can also be produced in HRCs using hybrid ("launch") vectors containing elements of both agrobacterial binary plasmids and plant viruses, as described by Musiychuk et al. [96]. For example, a hybrid vector pBI-D4 consists of sequences from an Agrobacterium binary plasmid, pBI121, and RNA of recombinant TMV [96]. These vector systems are widely used for transient expression of recombinant proteins, including vaccine antigens [97]. Hybrid vectors can also be stably integrated into the host plant's genome using an *A. rhizogenes*-mediated transformation, resulting in hairy root formation (Fig. 1). Target gene expression is derived from the continually replicating viral vector which, in turn, is constitutively transcribed from the nuclear genome. Moreover, because hairy roots are transgenic and each cell carries the vector, the need for virus cell-to-cell movement function is eliminated.

It is well documented that target expression depends on the ability of the host plant to sustain replication of the viral vector [98, 99]. Thus, in addition to engineering of an optimal expression construct, the choice of host plant species is important. The feasibility of long-term episomal protein production in stably transformed HRCs using the hybrid expression vector pBI-D4 has been demonstrated for several targets, including GFPC3 (Skarjinskaia et al. unpublished observation), lichenase (LicKM), a fusion product of LicKM to domain 4 of *Bacillus anthracis* protective antigen (PA) (current review), LicKMPA-D4 (current review) and a fusion product of LicKM to the E7 oncogene from human papillomavirus type 16 (HPV-16), LicKME7 [66].

Although hybrid vectors are designed to increase recombinant target expression, they are prone to vector instability. Therefore, inducible viral vectors combining the features of inducible promoters and plant viral sequences have been developed. Inducible plant viral vector systems include: an estradiol-inducible *tomato mosaic virus* (ToMV) amplicon expressing GFP in tobacco BY2 cells [100], an ethanol-inducible *bean yellow dwarf virus* (BeYDV) amplicon system expressing Norwalk virus capsid protein in tobacco NT-1 cells [101], an estradiol-inducible *cucumber mosaic virus* (CMV) viral amplicon, CMViva, for the production of functional recombinant proteins in nontransgenic leaf tissue using transient agroinfiltration [102, 103], the same system in transgenic tobacco suspension cultures [104], and an ethanol-inducible TMV viral replicon expressing GFP in stably transformed tobacco plants [85].

Collens et al. [105, 106] reported on a hybrid pGPTVK-GI vector consisting of an Agrobacterium binary vector, pBY031, and recombinant *beet yellow dwarf virus* (BYDV) containing the GUS reporter gene. HRCs of *Nicotiana glutinosa* were inoculated with *A. tumefaciens* bearing a hybrid vector [105]. Expression of GUS was recorded after 3 days post inoculation. Microscopic examination of histochemically stained root tissue revealed that GUS expression was mainly localized in actively dividing cells of root tips and the root zone of elongation. Such restricted localization of target leads to low overall expression levels of the reporter gene transiently delivered to hairy roots by a hybrid expression vector.

4 Disease Targets for Vaccine Candidates Produced in Hairy Root Cultures

Hepatitis B is an alarming viral disease, infecting about 2 billion people worldwide [107]. Currently produced hepatitis B vaccines comprise surface protein of the viral envelope, which also plays an important role in viral infection [107]. Proof of concept for a plant-produced hepatitis B vaccine was first established by generating transgenic tobacco plants expressing recombinant hepatitis B surface antigen (rHBsAg) in the form of virus-like particles (VLPs) analogous to those produced in yeast [108]. In a later study, potato HRCs expressing rHBsAg were generated [52]. Six hairy root lines were analyzed for growth and target antigen production in different culture media, demonstrating fivefold higher levels of HBsAg expression compared with the original transgenic plant (97.1 ng/g vs. 19.11 ng/g FW). No immunological study of hairy root-derived HBsAg has yet been reported.

Swine erysipelas is a disease caused by the bacterium *Erysipelothrix rhusiopathiae*, is associated with considerable economic loss, and continues to be a major problem in swine-producing areas worldwide [109]. The bacterium also causes a skin disease, erysipeloid, in humans. The repeat region at the C-terminus of immunogenic surface-protective antigen A of *E. rhusiopathiae*, SpaA, has been proposed as a candidate for a subunit vaccine against erysipelas [110]. SpaA expressed in *E. coli* completely protects mice against the *E. rhusiopathiae* challenge [110]. In an attempt to produce SpaA in plants, hairy roots were generated from transgenic tobacco plants designed to express SpaA fused to cholera toxin B subunit (CTB; [64]). However, when assessed by RT-PCR and Northern blot analysis, the emerged HRCs were found to express only low levels of SpaA.

ETEC is the leading cause of acute bacterial diarrhea in the developing world and the most common cause of travelers' diarrhea. Infection and colonization of the small intestine with ETEC can be fatal in the absence of treatment [111]. The heat-labile enterotoxin (LT) produced by ETEC consists of six subunits: a catalytically active A subunit (LTA) and a GM1 ganglioside binding B subunit pentamer (LTB; [112]). LTB was proposed as a mucosal immunogen [113]. In addition, it has been shown to be a potent mucosal adjuvant, stimulating antibody responses when co-expressed with vaccine antigens [114]. The development of the transgenic potato plants expressing LTB at up to 17 μg/g tuber has been reported by Mason et al. [22]. Recently, LTB has also been expressed in HRCs generated from *Solanaceae* plants including *Nicotiana tabacum*, *Lysopersicon esculentum*, and *Petunia parodii* [68]. Hairy roots were generated by stable leaf disk transformation with recombinant *A. rhizogenes* harboring a pBI121-based vector encoding LTB. The LTB expression level in a tobacco and petunia hairy roots comprised about 65–70 μg/g, whereas tomato hairy roots expressed only about 10 μg/g of LTB. Higher levels of LTB expression in HRCs caused lower levels of biomass accumulation, regardless of the plant species. The greatest growth retardation (80 %) relative to the control (hairy root lines transformed with a GUS-expressing construct) occurred in tobacco hairy roots and the lowest retardation (35 %) was observed in petunia hairy roots. LTB-expressing tomato HRCs did not reveal significant differences in the growth rate compared to the GUS control; however, LTB was produced in tomato hairy roots at an approximately six-times lower level than in petunia or tobacco hairy roots. Immunogenicity of LTB produced in HRCs was not assessed in this study.

Yersinia pestis, the causative agent of plague, is an extremely virulent bacterium, causing rapidly progressing disease and a high mortality rate. Fraction-1 capsular antigen (F1) and low calcium response virulent antigen (V) are immunogenic antigens of *Y. pestis* [115]. F1 protein has been suggested to inhibit phagocytosis of the pathogen by host macrophages [116]. V protein is secreted and inhibits an early host inflammatory response to infection [117]. Furthermore, in female BALB/c mice, an F1-V fusion protein produced in *E. coli* is protective when administered intramuscularly [118, 119]. In order to develop a mucosal vaccine against plague, an F1-V fusion protein was expressed in transgenic tomato plants [120]. It was shown that the percent of F1-V in total soluble protein in fully developed but still green tomato fruits was significantly higher ($P < 0.001$) than in ripe fruits and leaves. To increase shelf-life and standardize dose, transgenic green tomato fruits were freeze-dried, pooled, and powdered. BALB/c mice were first primed subcutaneously with the bacteria-produced F1-V fusion protein and then fed with lyophilized transgenic tomato powder. The F1- and V-specific serum IgG1 antibodies were detected in 100 % of immunized animals. In addition, F1- and V-specific mucosal IgA antibodies were detected in fecal pellets of the immunized mice [120]. The F1-V fusion protein of *Y. pestis* was also produced in transgenic tobacco plants and hairy roots [65]. Expression was driven by the constitutive *ocs-mas* superpromoter [73] fused to *tobacco etch virus* translational enhancer [121]. For subcellular targeting of the fusion protein, a signal peptide from the potato patatin gene (*Pat*) was included at the N-terminus of the target. A hexa-histidine tag (6xHis) was also included to facilitate detection and purification of the target. Additionally, the expression construct contained the lectin subunit of ricin, ricin B (RTB), which is a potent mucosal adjuvant and an antigen carrier for oral immunization [17]. Expression analysis of the RTB:F1:V antigen fusion revealed the lowest expression in leaves and the highest in transgenic plant roots and hairy roots derived from transgenic plants, suggesting that HRCs may be suitable for recovering large quantities of the target protein for use as a plague vaccine candidate. The 6xHis tag-labeled RTB:F1:V fusion protein was purified from HRCs for evaluation of its ability to elicit systemic and mucosal antibody responses following nasal immunization of mice. However, there are no reports on immunogenicity of RTB:F1:V produced in HRCs.

HPV causes the second highest incidence of death through cancer worldwide in women [122]. The HPV-16 type of the virus is recognized as a causative agent of cervical cancer [123]. The genome of HPV consists of six nonstructural viral regulatory E-proteins and two structural viral capsid L-proteins [124]. The E6 and E7 oncoproteins are responsible for malignant transformation of host cells by down-regulation of tumor-suppressing genes, which has been suggested as a mechanism of tumor induction by HPV [125]. As a candidate vaccine against cervical cancer, E7 oncoprotein from the high-risk HPV-16 strain has been produced in plants. Agrobacterium infiltration of *N. benthamiana* with the pBI-D4 launch vector encoding HPV-16 E7 fused to the LicKM carrier resulted in transient expression of the LicKM-HPV-16 E7 fusion that was purified and evaluated in a mouse model for its potential as a prophylactic and therapeutic vaccine [126]. The vaccine was shown to be immunogenic, inducing E7-specific IgG antibody and cytotoxic T cell responses, and protected mice against lethal challenge with an E7-expressing tumor cell line. Massa et al. [66] also generated HRCs from *N. benthamiana* and *Petunia hybrida* by stable transformation with recombinant *A. rhizogenes* harboring a hybrid pBI-D4-based vector including sequence encoding the LicKM-HPV-16 E7 antigen fusion. The authors indicated an intention to produce hairy root extracts and lyophilized material for pre-clinical evaluation, but results have not been reported.

Crimean–Congo hemorrhagic fever (CCHF) is a human tick-borne viral disease. CCHF is frequently fatal, and the current approach to the disease management comprises general supportive measures and administering an antiviral drug, ribavirin [127]. There is no safe and effective vaccine against CCHF, although an inactivated, mouse-brain–derived vaccine has been developed and used for prophylaxis of defined human risk groups [128]. An attempt to produce a plant-made vaccine against CCHF based on the immunogenic properties of CCHF virus (CCHFV) glycoproteins was reported by Ghiasi et al. [129]. Viruses of this family, *Bunyaviridae*, contain three structural proteins: two envelope glycoproteins (Gc and Gn), a nucleocapsid protein, and a RNA-dependent RNA polymerase [130]. The viral glycoproteins are recognized by the receptor sites on susceptible host cells. Viruses are internalized by endocytosis and replicate in the cytoplasm of infected cells [130]. The immunogenicity of Gc and Gn glycoproteins was demonstrated in mice immunized with an experimental DNA vaccine for CCHFV [131]. A recombinant G1/G2 (Gc/Gn) glycoprotein of CCHFV has been produced in leaves and hairy roots of transgenic tobacco and tested for immunogenicity in an animal model [129]. The G1/G2 glycoprotein was expressed at 1.2 μg/g FW of transgenic leaves and 1.8 μg/g FW of hairy roots. BALB/c mice were fed with either pelleted transgenic hairy roots or leaves, and a subset of these mice was then also injected with a single low dose of purified G1/G2 as a booster. Positive control mice were immunized with the current CCHF vaccine. Both transgenic leaf and hairy root material containing recombinant G1/G2 were shown to be similarly immunogenic when delivered orally to mice, eliciting anti-G1/G2 serum IgG and fecal IgA antibodies in 100 % of immunized animals. The boosted groups showed a large increase in anti-G1/G2 IgG antibodies. Although the G1/G2 glycoprotein from transgenic leaves and hairy roots had similar immunogenicity, the authors suggested that the hairy root system is more promising compared with the transgenic tobacco leaf system, due to potential for large-scale cultivation in culture medium and faster growth.

Bacillus anthracis is one of the potential agents of most concern for biological warfare. The main virulence factor of *B. anthracis* is a lethal toxin consisting of PA and either lethal factor (LF) or edema factor (EF). PA is recognized as the most effective immunogen for vaccination against anthrax [132]. Aziz et al. [133] first employed the full-length PA protein (83 kDa) to generate an oral vaccine against anthrax by expressing the antigen in transgenic tobacco. The functionality of plant-produced PA was assessed in a cytolytic assay, demonstrating that in combination with LF it was able to lyse macrophages and had a biological activity comparable to that of native PA. In a subsequent study, immunization of mice with either recombinant PA or domain 4 of PA (PAD4), or domain 1 of LF (LFD1) conferred protection against anthrax infection [134]. The importance of PAD4 was illustrated by the fact that deletion of the sequence encoding this domain from the *B. anthracis* genome leads to a > 5,000-fold decrease in virulence [135].

We have transiently expressed PAD4 and LFD1, both as stand-alone proteins and as fusions with the LicKM carrier molecule, in *N. benthamiana* (Fig. 2). LicKM and LicKM–PAD4 expression cassettes were cloned into the pBI-D4

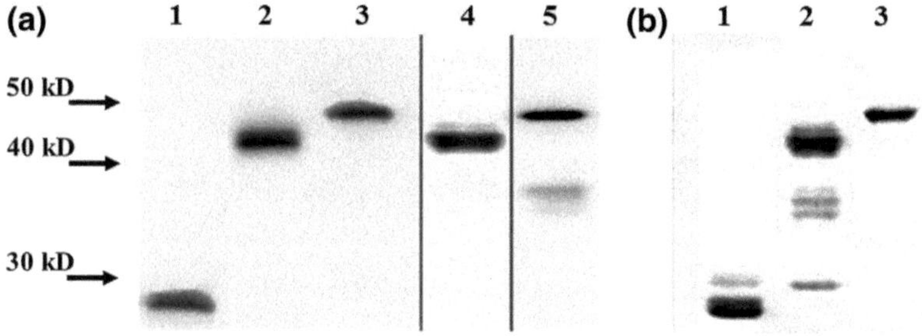

Fig. 2 **a** Western blot analysis of purified plant-produced LicKM (1), LicKM–PAD4 (2), and LicKM–LFD1 (3) probed with an anti-LicKM polyclonal antibody; LicKM–PAD4 probed with an anti-PA monoclonal antibody (4), and LicKM–LFD1 probed with an anti-LF antibody (5). **b** Coomassie-stained gel with purified plant-produced LicKM (1), LicKM–PAD4 (2), and LicKM–LFD1 (3)

vector [96], and the resulting constructs were introduced into *A. rhizogenes* strain A4RSII, followed by infiltration into *N. benthamiana*. Transient expression of LicKM–LFD1 was achieved by inoculating *N. benthamiana* leaves with an in vitro transcript of the viral vector, 30B-LicKM–LFD1 [136].

The LicKM–PAD4 gene fusion was also stably integrated into the genome of *Petunia hybrida*, a low alkaloid variant (kindly provided by Dr. John Hammond of the US Department of Agriculture) using the pBI-D4 hybrid vector. *P. hybrida* sustains TMV replication and movement within plant tissue. Furthermore, the plant does not produce toxic alkaloids which might be harmful to experimental animals consuming petunia hairy roots. Petunia hairy roots were generated by *A. rhizogenes*-mediated transformation, selected and further propagated on K3 culture medium [48]. The best expressing hairy root clones were identified at early stages of growth using a lichenase-specific colorimetric enzymatic assay [90] using 2–3 mg of root tissue. Hairy root clones that demonstrated the highest lichenase activity were propagated further. SDS-PAGE followed by immunoblot analysis using a human anti-PA monoclonal antibody (IQ Corporation, The Netherlands) or a rabbit polyclonal anti-LicKM antibody (Washington Biotechnology Inc., USA) were performed using 20–50 mg of root tissue [48]. LicKM–PAD4 and LicKM–LFD1 were purified from *N. benthamiana* plants [96, 136] and quantified against a bovine serum albumin (BSA) standard. Plant-produced purified LicKM–PAD4 of a defined concentration was used as a protein standard in Western blot analysis.

A total of 204 clonal hairy root lines were screened to determine the expression level of LicKM–PAD4 and root biomass accumulation during the standard subculture period (4 weeks). Clonal line #26, with an expression level of about 8 mg/kg of blotted fresh weight (FW; Fig. 3), was propagated in 2-L glass medicinal jars by subculturing a portion of the biomass in fresh medium for over 2 years. At 6-week intervals, root biomass was harvested and the FW was determined. The increase in root biomass during subculturing was more than 30-fold. Blotted roots from each of five culture jars were combined and freeze-dried in a lyophilizer

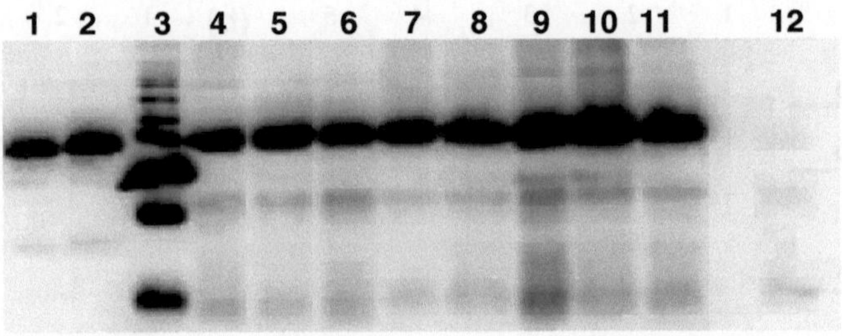

Fig. 3 Western blot analysis of LicKM–PAD4 expression in clonal hairy root lines. LicKM–PAD4 expression was quantified against a plant-produced standard protein, LicKM–PAD4. Root extracts were probed with an anti-PA monoclonal antibody. *Lanes* 1 and 2: 5 and 10 ng of the standard protein, respectively. *Lane* 3: molecular weight marker. *Lanes* 4–11: total protein from hairy root extracts. Each lane represents a different clonal hairy root line. *Lane* 10 represents line #26 which was chosen as a production hairy root line with the expression level of 8 mg/kg of blotted FW. *Lane* 12: negative control, an extract from hairy roots expressing LicKM. Protein extract loaded in each well corresponds to 2 mg of root tissue

(FreeZone 4.5, Labconco). Lyophilized tissue was stored at –80 °C. To prepare lyophilized material for oral immunization, the powdered content of multiple bags was combined and thoroughly mixed to achieve homogeneity of the expressed antigen confirmed by Western blot analysis using an anti-PA antibody (Fig. 4). During lyophilization, the tissue weight loss was 13-fold (FW to dry weight). To reconstitute FW, 10 mg of lyophilized root powder were re-suspended in 130 μL of extraction buffer. Expression of the antigen in nine random samples of lyophilized tissue comprised 4.10, 5.32, 5.34, 7.24, 5.82, 6.2, 6.28, 5.89, and 5.80 ng (lanes 5–13, respectively), with the mean value of 5.78 ± 0.85 ng of LicKM–PAD4 (Fig. 4). Based on these results, the concentration of LicKM–PAD4 in the lyophilized root tissue prepared for oral immunization was uniform and comprised ~38 mg of active LicKM–PAD4 per kg of dried petunia hairy root biomass.

To assess the biological activity of LicKM–PAD4 produced in petunia hairy roots, the lyophilized hairy root material containing LicKM–PAD4 was used to mucosally boost LicKM–PAD4-primed mice. Mice were initially immunized intraperitoneally at 2-week intervals with three doses of 100 μg of plant-produced LicKM–PAD4 with or without 100 μg of LicKM–LFD1. Ten weeks after the last injection, each animal received 30 mg of dried root material reconstituted in water by oral gavage, comprising a dose of ~1 μg of LicKM–PAD4 per administration. The mucosal boost was repeated after 1 week. Post second boost, LicKM–PAD4-immunized mice had 0.34 mg/mL of an anti-PA IgG serum antibody, and mice immunized with LicKM–PAD4/LFD1 had 2.3 mg/mL of an anti-PA IgG serum antibody (Fig. 5). Mice in the LicKM negative control group had negligible anti-PA antibody titers. As expected, mice immunized with full-length PA generated the highest antibody titers, with a serum concentration of anti-PA IgG of 1 mg/mL even following the priming dose, which increased to 80 mg/mL after the second boost. These results indicate that LicKM–PAD4 and LicKM–LFD1 fusion proteins produced in plants can elicit strong humoral immune responses in mice. Furthermore,

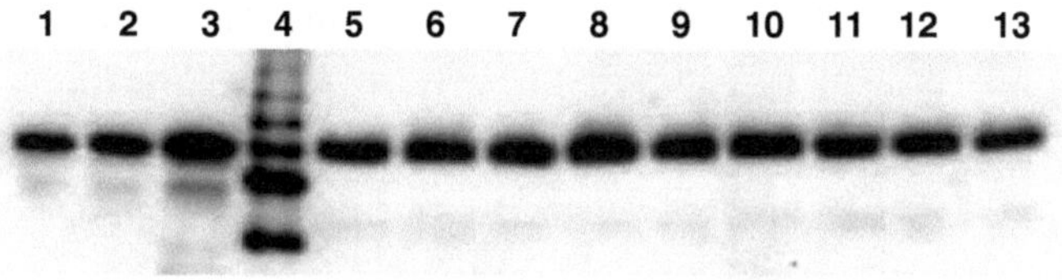

Fig. 4 Concentration of the LicKM–PAD4 antigen in lyophilized bulked root tissue prepared for oral immunization. Expression was quantified against a plant-produced standard protein, LicKM–PAD4. Root extracts were probed with an anti-PA monoclonal antibody. *Lanes* 1–3: 2.5, 5, and 10 ng of protein standard, respectively. *Lane* 4: molecular weight marker. *Lanes* 5–13: expression of the antigen in nine random samples of lyophilized tissue. Ten μL of protein extract were loaded in each well, corresponding to 2 mg of reconstituted FW of root tissue

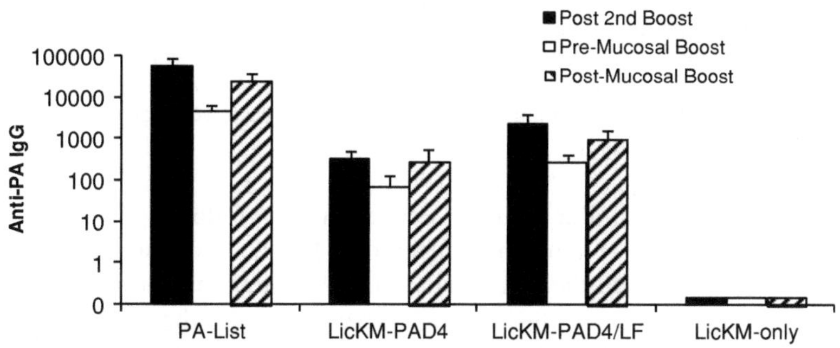

Fig. 5 Mucosal administration of root material expressing LicKM–PAD4 boosts IgG antibody responses in PA-primed mice. Groups of six female A/J mice were injected intraperitoneally at 2-week intervals with three doses of 100 μg of purified plant-produced LicKM–PAD4 with or without 100 μg of LicKM–LFD1, emulsified in Freund's adjuvant (Sigma-Aldrich, St. Louis, MO). Control animals received 50 μg of LicKM or recombinant full-length *B. anthracis* PA (List Biological Laboratories Inc.). Ten weeks after the last injection, mice were mucosally boosted two times at a 1-week interval with the lyophilized petunia hairy root material containing ~38 mg/kg of LicKM–PAD4 by oral gavage for a dose of ~1 μg of LicKM–PAD4 per boost. Serum samples were collected prior to each immunization and 1 week after the last mucosal boost and analyzed using an anti-PA IgG enzyme-linked immunosorbent assay [136]. The amount of PA-specific IgG was calculated based on standard curves prepared using a mouse anti-PA IgG serum standard (kindly provided by Susan Wimer-Mackin of Ligocyte Pharmaceuticals, Inc., Bozeman, MT). The results are expressed as μg of anti-PA IgG per mL of serum

the addition of LicKM–LFD1 to LicKM–PAD4 significantly increased the amount of the anti-PA IgG antibody in the serum. An increase in the serum anti-PA antibody concentrations in test groups after the second mucosal boost supports the feasibility of using the transgenic hairy root material to mucosally boost prior humoral response.

The stability of the LicKM–PAD4 antigen in petunia HRCs was also assessed. Samples of lyophilized powdered root tissue were prepared and stored at ambient temperature or at –80 °C. Western blot analysis of LicKM–PAD4 was performed following storage of these samples for 14 months after the sample preparation. The

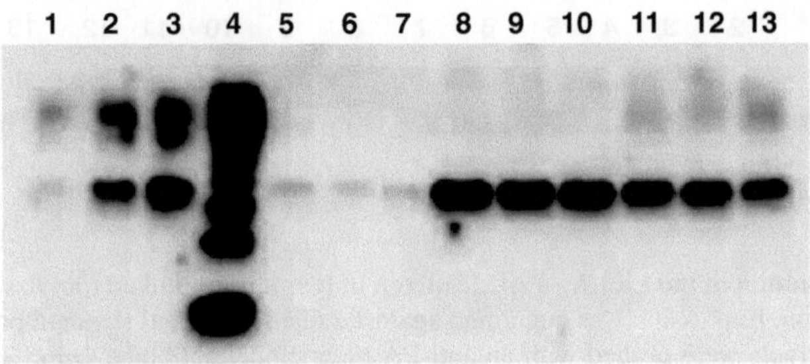

Fig. 6 Stability of LicKM–PAD4 in lyophilized root tissue stored at –80 °C for 14 months. LicKM–PAD4 levels were quantified against a plant-produced standard, LicKM–PAD4. Root extracts were probed with an anti-PA monoclonal antibody. *Lanes* 1–3: 1, 2.5, and 5 ng of the standard, respectively. *Lane* 4: molecular weight marker. *Lanes* 5–7: expression of the antigen in protein extracts prepared from lyophilized tissue and stored at –80 °C. *Lanes* 8–10: expression of the antigen in protein extracts prepared from the lyophilized tissue stored at –80 °C. *Lanes* 11–13: expression of antigen in protein extracts prepared from the lyophilized tissue stored at ambient temperature. Ten μL of protein extract were loaded in each well, corresponding to 2 mg of reconstituted FW of root tissue

results demonstrated similar antigen concentrations following storage of the lyophilized root tissue at ambient temperature [3.43, 3.46, and 2.40 ng (lanes 11–13, respectively), with the mean value of 3.09 ± 0.60 ng] or at –80 °C [5.42, 5.27, and 4.89 ng (lanes 8–10, respectively), with the mean value of 5.19 ± 0.27 ng; Fig. 6]. In contrast, extracts prepared from the lyophilized tissue and stored at –80 °C showed a dramatic decline in PAD4 [0.30, 0.21, and 0.26 ng (lanes 5–7, respectively), with the mean value of 0.26 ± 0.04 ng; Fig. 6]. The slight decrease in the concentration of PAD4 in lyophilized samples stored at ambient temperature may be due to the storage conditions. Although sampled tissue was stored in tightly closed tubes, some humidification of lyophilized tissue may have occurred, leading to a decline in the antigen concentration. Thus, for long-term storage, additional precautions, such as storage of samples with silica gel or in vacuum-sealed containers, should be considered.

5 Conclusions

Plants offer significant opportunities for making safe and effective subunit vaccines. The first plant-produced vaccine candidate was against hepatitis B and was reported in the 1980s. Since then, many transgenic plant lines expressing antigen targets of viral, bacterial, or parasitic origin have been generated, tested in animals, and demonstrated to be immunogenic, and in some cases efficacious. In recent years, the HRC-based production of target proteins has been reported, and the hairy root-produced proteins had the desired biological activity. Hairy roots can be used for producing proteins in a fully contained environment, eliminating the majority of concerns associated with the use of transgenic plants. Despite these advantages, the hairy root platform still requires significant improvements to become commercially feasible. In particular, it is essential to enhance productivity by careful genotype selection and molecular engineering of expression constructs, as well as to design bioreactors suited for optimal hairy root cell proliferation and target protein production.

Acknowledgments The authors would like to thank Ms. Margaret Shillingford for assistance with greenhouse-grown plants and Dr. Natasha Kushnir for editorial assistance.

References

1. Rai M, Padh H (2001) Expression systems for production of heterologous proteins. Curr Sci 80:1121–1128
2. Twyman RM, Schillberg S, Fischer R (2005) Transgenic plants in the biopharmaceutical market. Expert Opin Emerg Drugs 10:185–218
3. Sharma A, Sharma MK (2009) Plants as bioreactors: Recent developments and emerging opportunities. Biotechnol Adv 27:811–832
4. Thomas BR, Van Deynze AE, Bradford KJ (2002) Production of therapeutic proteins in plants. Agric Biotechnol California Series, Publication 8078. http://anrcatalog.ucdavis.edu/
5. Faye L, Gomord V (2010) Success stories in molecular farming—a brief overview. Plant Biotechnol J Special Issue: Success Stories in Molecular Farming 8:525–528
6. Rigano M, Walmsley A (2005) Expression systems and developments in plant-made vaccines. Immunol Cell Biol 83:271–277
7. Neutra MR, Kozlowski PA (2006) Mucosal vaccines: the promise and the challenge. Nature Rev Immunol 6:148–158
8. Kaul D, Ogra PL (1998) Mucosal responses to parenteral and mucosal vaccines. Dev Biol Stand 95:141–146
9. Gomes E, Zoth S, Carillo E et al (2010) Developments in plant-based vaccines against diseases of concern in developing countries. The Open Infect Dis J 4:55–62
10. Freytag LC, Clements JD (1999) Bacterial toxins as mucosal adjuvants. Curr Topics Microbiol Immunol 236:215–236
11. Landridge WHR (2000) Edible vaccines. Sci Am 283:48–53
12. Pizza M, Guiliani M, Fontana E et al (2001) Mucosal vaccines: non-toxic derivatives of LT and CT as mucosal adjuvants. Vaccine 19:2534–2541
13. Rhee JH, Lee SE, Kim SY (2012) Mucosal vaccine adjuvants update. Clin Exp Res 1:50–63
14. Clements JD, Hartzog NM, Lyon FL (1988) Adjuvant activity of *Escherichia coli* heat-labile enterotoxin and effect on the induction of oral tolerance in mice to unrelated protein antigens. Vaccine 6:269–277
15. Elson CJ, Ealding W (1984) Generalised systemic and mucosal immunity in mice after mucosal stimulation with cholera toxin. J Immunol 132:2736–2741
16. Lavelle EC, Grant G, Pusztai A et al (2001) The identification of plant lectins with mucosal adjuvant activity. Immunology 102:77–86
17. Medina-Bolivar F, Wright R, Funk V et al (2003) A non-toxic lectin for antigen delivery of plant-based mucosal vaccines. Vaccine 21:997–1005
18. Streatfield S (2006) Mucosal immunization using recombinant plant-based oral vaccines. Methods 38:150–157
19. Aliahmadi A, Rahmani N, Abdollahi M (2006) Plant-derived human vaccines: an overview. Int J Pharmacol 2:268–279

20. Mishra N, Gupta PN, Khatri K et al (2008) Edible vaccines: A new approach to oral immunization. Indian J Biotechnol 7:283–294
21. Ahmad P, Ashraf M, Younis M et al (2012) Role of transgenic plants in agriculture and biopharming. Biotech Advances 33:524–540
22. Mason HS, Hag TA, Clements JD et al (1998) Edible vaccine protects mice against *Escherichia coli* heat-labile enterotoxin (LT): potatoes expressing a synthetic LT-B gene. Vaccine 16:1336–1343
23. Marquet-Blouin E, Bouche FB, Steinmetz A et al (2003) Neutralizing immunogeneicity of transgenic carrot (*Daucus carota L.*)-derived measles virus hemagglutinin. Plant Mol Biol 51:459–469
24. Chowdhury K, Bagasra O (2007) An edible vaccine for malaria using transgenic tomatoes of varying sizes, shapes and colors to carry different antigens. Med Hypotheses 68:22–30
25. Arakawa T, Chong KX, Merritt JL et al (1997) Expression of cholera toxin B subunit oligomers in transgenic potato plants. Transgenic Res 6:403–410
26. Wigdorovitz A, Carrillo C, Dos Santos MJ (1999) Induction of a protective antibody response to foot and mouse disease virus in mice following oral or parenteral immunization with alfalfa transgenic plants expressing the viral structural protein VP1. Virology 255:347–353
27. Richter L, Thanavala Y, Arntzen C et al (2000) Production of hepatitis B surface antigen in transgenic plants for oral immunization. Nat Biotechnol 18:1167–1171
28. Pniewski T, Kapusta J, Bociag P et al (2011) Low-dose oral immunization with lyophilized tissue of herbicide-resistant lettuce expressing hepatitis B surface antigen for prototype plant-derived vaccine tablet formulation. J Appl Genet 52:125–136
29. Kapusta J, Modelska A, Figlerowicz M et al (1999) A plant derived edible vaccine against hepatitis B virus. FASEB 13:1796–1799
30. Tacket CO, Pasetti MF, Edelman R (2004) Immunogenicity of recombinant LT-B delivered orally to humans in transgenic corn. Vaccine 22:4385–4389
31. Thanavala Y, Mahoney M, Pal S (2005) Immunogenicity in humans of an edible vaccine for hepatitis B. Proc Natl Acad Sci USA 102:3378–3382
32. Yusibov V, Hooper DC, Spitsin SV et al (2002) Expression in plants and immunogenicity of plant virus-based experimental rabies vaccine. Vaccine 20:3155–3164
33. Walmsley A, Arntzen C (2003) Plant cell factories and mucosal vaccines. Curr Opin Biotechnol 14:145–150
34. Schillberg S, Twyman R, Fischer R (2005) Opportunities for recombinant antigen and antibody expression in transgenic plants-biotechnology assessment. Vaccine 23:1764–1769
35. Guillon S, Trémeouillax-Guiller J, Pati PK et al (2006) Hairy root research: Recent scenario and exciting prospects. Curr Opin Plant Biol 9:341–346
36. Ronchi VN, Martini G, Buiatti M (1975) Genotype-hormone interaction in the induction of chromosome aberrations: Effect of 2.4-dichlorophenoxyacetic acid (2,4-D) and kinetin on tissue cultures from nicotiana SPP. Mutat Res Fund Mol Mech Mut 36:67–72
37. Phillips R, Kaepplert S, Olhoft P (1994) Genetic instability of plant tissue cultures: breakdown of normal controls (DNA methylation/genome rearrangements/rearrangement induced premeiotically/repeat-induced point mutation/heterochromatin). Proc Natl Acad Sci USA 91:5222–5226
38. Sevon N, Oksman-Caldentey K-M (2002) *Agrobacterium rhizogenes* mediated transformation: root cultures as a source of alkaloids. Planta Med 68:859–868
39. Flores H, Vivanco J, Loyola-Vargas V (1999) "Radicle" biochemistry: the biology of root-specific metabolism. Trends Plant Sci 4:220–226
40. Shanks J, Morgan J (1999) Plant 'hairy root' culture. Curr Opin Biotechnol 10:151–155
41. Rao SR, Ravishankar GA (2002) Plant cell cultures: Chemical factories of secondary methabolites. Biotechnol Adv 20:101–153
42. Medina-Bolivar F, Condori J, Rimando AM et al (2007) Production and secretion of resveratrol in hairy root cultures of peanut. Phytochemistry 68:1992–2003

43. Srivastava S, Srivastava A (2007) Hairy root culture for mass-production of high-value secondary metabolites. Crit Rev Biotechnol 27:29–43
44. Hu ZB, Du M (2006) Hairy root and its application in plant genetic engineering. J Integr Plant Biol 48:121–127
45. Doran P (2002) Properties and applications of hairy root cultures. In: Oksman-Caldentey K-M, Barz WH (eds) Plant Biotechnology and Transgenic Plants. Marcel Dekker Inc, New York
46. Wongsamuth R, Doran P (1997) Production of monoclonal antibodies by tobacco hairy roots. Biotechnol Bioeng 54:401–415
47. Cooke DE, Webb KJ (1997) Stability of 35S gus gene expression in (birds foot trefoil) hairy root cultures under different growth conditions. Plant Cell Tiss Org Cult 47:163–168
48. Skarjinskaia M, Karl J, Araujo A et al (2008) Production of recombinant proteins in clonal root cultures using episomal expression vectors. Biotechnol Bioeng 100:814–819
49. Marconi PL, Setten LM, Calcena EN et al (2008) Changes in growth and tropane alkaloid production in long-term culture of hairy roots of *Brugmansia candida.* J Integr Biosci 3:38–44
50. Baiza AM, Quiros-Moreno A, Ruiz JA (1999) Genetic stability of hairy root cultures of *Datura stramonium.* Plant Cell Tiss Org Cult 59:9–17
51. Woods R, Geyer B, Mor T (2008) Hairy-root organ cultures for the production of human acetylcholinesterase. BMC Biotechnol 8:95–101
52. Sunil Kumar GB, Ganapathi TR, Srinivas L (2006) Expression of hepatitis B surface antigen in potato hairy roots. Plant Sci 170:918–925
53. Maldonaldo-Mendoza IE, Ayora-Talavera T, Loyola-Vargas VM (1993) Establishment of hairy root cultures of *Datura stramonium.* Characterization and stability of tropane alkaloid production during long period of sub-culturing. Plant Cell Tiss Org Cult 33:321–329
54. Palazón J, Moyano E, Cusidó RM (2003) Alkaloid production in *Duboisia* hybrid hairy root and plants overexpressing the *h6h* gene. Plant Sci 165:1289–1295
55. Menzel G, Harloff HJ, Jung C (2003) Expression of bacterial poly (3-hydroxybutyrate) synthesis genes in hairy roots of sugar beet (*Beta vulgaris L.*). Appl Microbiol Biotechnol 60:571–576
56. Moyano E, Jouhikainen K, Tammela P (2003) Effect of *pmt* gene overexpression on tropane alkaloid production in transformed root cultures of *Datura metel* and *Hyoscyamus muticus.* J Exp Bot 54:203–211
57. Drake PMW, Chargelegue DM, Vine ND et al (2003) Rhizosecretion of a monoclonal antibody protein complex from transgenic tobacco roots. Plant Mol Biol 52:233–241
58. Pham NG, Schafizrl H, Wink M (2012) Production and secretion of recombinant thaumatin in the hairy root cultures. Biotechnol J 7:537–545
59. Giri A, Narasu ML (2000) Transgenic hairy roots: recent trends and applications. Biotechnol Adv 18:1–22
60. Sharp JM, Doran PM (1999) Effect of bacitracin on growth and monoclonal antibody production by tobacco hairy roots and cell suspensions. Biotechnol Bioprocess Eng 4:253–258
61. Terashima M, Murai Y, Kawamura M (1999) Production of functional human alpha 1-antitrypsin by plant cell culture. Appl Microbiol Biotechnol 52:516–523
62. Hellwig S, Drossard J, Twyman R et al (2004) Plant cell cultures for the production of recombinant proteins. Nature Biotechnol 22:1415–1422
63. Martinez C, Petrucelli S, Giulietti AM et al (2005) Expression of the antibody 14D9 in *Nicotiana tabacum* hairy roots. Elect J Biotechn 8:170–176
64. Ko S, Liu J-R, Yamakawa T et al (2006) Expression of the protective antigen (SpaA) in transgenic hairy roots of tobacco. Plant Mol Biol Reporter 24:251
65. Woffenden BJ, Nopo LH, Cramer CL (2008) Expression of a ricin B:F1:V fusion protein in tobacco hairy roots: steps toward a novel pneumonic plaque vaccine. Elect J Integr Biosci 3:10–19
66. Massa S, Skarjinskaia M, Mett V et al (2009) Plant platforms for producing anti-cancer therapeutic vaccines. Plant-Based Vaccines & Antibodies Plant Expression Systems for Recombinant Pharmacologics PBVA Verona, Italy, Presented at

67. Demurtas O, Franconi R, Massa S (2010) Plant-derived vaccines and other therapeutics produced in contained systems. Expert Rev Vaccines 9:877–898
68. Guzman G, Walmsley A, Webster D et al (2011) Hairy roots cultures from different Solanaceous species have varying capacities to produce *E. coli* B-subunit heat-labile toxin antigen. Biotechnol Lett 33:2495–2502
69. Simpson RB, Spielmann A, Margossian L et al (1986) A disarmed binary vector from Agrobacterium tumefaciens functions in Agrobacterium rhizogenes. Plant Mol Biol 6:403–415
70. Gelvin SB (2000) Agrobacterium and plant genes involved in T-DNA transfer and integration. Annu Rev Plant Phys Plant Mol Biol 51:223–256
71. Streatfield S (2007) Approaches to achieve high-level heterologous protein production in plants. Plant Biotechnol 5:2–15
72. Odell J, Nagy F, Chua N (1985) Identification of DNA sequences required for activity of the cauliflower virus 35S promoter. Nature 313:810–812
73. Lee L-Y, Kononov M, Bassuner B et al (2007) Novel Plant Transformation Vector Containing the Super-promoter. Plant Physiol 145:1294–1300
74. Jones M, Manning K, Andrews J et al (2008) The promoter from *SlREO*, a highly-expressed, root-specific *Solanum lycopersicum* gene, directs expression to cortex of mature roots. Funct Plant Biol 35:1224–1233
75. Dujovny G, Valli A, Calvo M et al (2009) A temperature-controlled amplicon system derived from Plum pox potyvirus. Plant Biotechnol J 7:49–58
76. Maiti I, Wagner G, Hunt A (1991) Light inducible and tissue-specific expression of a chimeric mouse metallothionein cDNA gene in tobacco. Plant Sci 76:99–107
77. Guzman G, Walmsley A, Webster D et al (2012) Use of the wound-inducible NtQPT2 promoter from *Nicotiana tabacum* for production of a plant-made vaccine. Biotechnol Lett 34:1143–1150
78. Mett V, Lochhead L, Reynolds P (1993) Copper-controllable gene expression system for whole plants (transcription activation/yeast metalloregulatory transcription factor/chimeric promoter). Proc Natl Acad Sci USA 90:4567–4571
79. Caddick M, Greenland A, Jepson I et al (1998) An ethanol inducible gene switch for plants used to manipulate carbon metabolism. Nat Biotechnol 16:177–180
80. Sweetman J, Chu C, Qu N (2002) Ethanol vapor is an efficient inducer of the *alc* gene expression system in model and crop plant species. Plant Physiol 129:943–948
81. Schena M, Lloyd AM, Dawis RD (1991) A steroid-inducible gene expression system for plant cells. Proc Natl Acad Sci USA 88:10421–10425
82. De Veylder L, van Montagu M, Inze D (1997) Herbicide safener-inducible gene expression in *Arabidopsis thaliana*. Plant Cell Physiol 38:568–577
83. Martinez A, Sparks C, Hart CA (1999) Ecdysone agonist inducible transcription in transgenic tobacco plants. Plant J 5:559–569
84. Padidam M, Gore M, Lu L et al (2003) Chemical-inducible, ecdysone receptor-based gene expression system for plants. Transgen Res 12:101–109
85. Werner S, Breus O, Symonenko Y et al (2011) High-level recombinant protein expression in transgenic plants by using a double-inducible viral vector. Proc Natl Acad Sci USA 108:14061–14066
86. Gallie D, Walbot V (1992) Identification of the motifs within the tobacco mosaic virus 5' leader responsible for enhancing translation. Nucleic Acids Res 20:4631–4638
87. Laguia-Becher M, Martin V, Kraemer M et al (2010) Effect of codon optimization and subcellular targeting on *Toxoplasma gondii* antigen SAG1 expression in tobacco leaves to use in subcutaneous and oral immunization in mice. BMC Biotechnol 10:52–65
88. Nuttal J, Vine N, Hadlington JL (2002) ER-resident chaperone interactions with recombinant antibodies in transgenic plants. Eur J Biochem 269:6042–6051
89. Gomord V, Faye L (2004) Posttranslational modification of therapeutic proteins in plants. Curr Opin Plant Biol 7:171–181

90. Musiychuk KA, Goldenkova IV, Avdeev RM et al (2000) Preparation and properties of Clostridium thermocellum lichenase deletion variants and their use for construction of bifunctional hybrid proteins. Biochemistry (Mosc) 65:1397–1402
91. Hull R (1989) The movement of viruses in plants. Annu Rev Phytopathol 27:213–240
92. Dawson WO, Bubrick P, Grantham GL (1988) Modifications of the tobacco mosaic virus coat protein gene affecting replication, movement and symptomatology. Phytopathology 78:783–789
93. Shadwick FS, Doran P (2007) Propagation of plant viruses in hairy root cultures: a potential method for in vitro production of epitope vaccines and foreign proteins. Biotechnol Bioeng 96:570–583
94. Shivprasad S, Pogue G, Lewandowski D et al (1999) Heterologous sequences greatly affect foreign gene expression in tobacco mosaic virus-based vectors. Virology 255:312–323
95. Rabindran S, Dawson W (2001) Assessment of recombinants that arise from the use of a TMV-based transient expression vector. Virology 284:182–189
96. Musiychuk K, Stephenson N, Bi H et al (2007) A launch vector for the production of vaccine antigens in plants. Influenza Other Respi Viruses 1:19–25
97. Yusibov V, Streatfield SJ, Kushnir N et al (In Press) Hybrid viral vectors for vaccine and antibody production in plants. Curr Pharm Design
98. Dawson WO, Hilf ME (1992) Host-range determinants of plant viruses. Annu Rev Plant Physiol Plant Mol Biol 43:527–555
99. Schaad MC, Carrington JC (1996) Suppression of long-distance movement of tobacco etch virus in a nonsusceptible host. J Virol 70:2556–2561
100. Dohi K, Nishikiori M, Tamai A et al (2006) Inducible virus-mediated expression of a foreign protein in suspension-cultures cells. Arch Virol 151:1075–1084
101. Zhang X, Mason H (2006) Bean Yellow Dwarf Virus replicons for high-level transgene expression in transgenic plants and cell cultures. Biotechnol Bioeng 93:271–279
102. Sudarshana MR, Plesha MA, Uratsu SL et al (2006) A chemically inducible cucumber mosaic virus amplicon system for expression of heterologous proteins in plant tissues. Plant Biotechnol J 4:551–559
103. Plesha MA, Huang T-K, Dandekar AM et al (2009) Optimization of the bioprocessing conditions for scale-up of transient production of a heterologous protein in plants using a chemically inducible viral amplicon expression system. Biotechnol Prog 25:722–734
104. Huang TK, Plesha MA, Falk BW et al (2009) Bioreactor strategies for improving production yield and functionality of a recombinant human protein in transgenic tobacco cell cultures. Biotechnol Bioeng 102:508–520
105. Collens JL, Lee DR, Seeman AM et al (2004) Development of auxotrophic agrobacterium tumefaciens for gene transfer in plant tissue culture. Biotechnol Prog 20:890–896
106. Collens JI, Mason HS, Curtis WR (2007) Agrobacterium-mediated viral vector-amplified transient gene expression in *Nicotiana glutinosa* plant tissue culture. Biotechnol Prog 23:570–576
107. WHO. World Hepatitis Day. 28 July 2012. www.who.int/csr/disease/hepatitis/world_hepatitis_day/en/ (accessed 17 September 2012)
108. Mason H, Lam D, Arntzen C (1992) Expression of hepatitis B surface antigen in transgenic plants. Proc Natl Acad Sci USA 89:11745–11749
109. Wood RL (1984) Swine erysipelas—a review of prevalence and research. J Am Vet Med Assoc 184:944–948
110. Makino S-I, Yamamoto K, Murakami S et al (1998) Properties of repeat domain found in a novel protective antigen, SpaA, of Erysiopelothrix rhusiopathiae. Microb Pathog 723:101–109
111. WHO. Diarrheal Diseases (Updated February 2009) www.who.int/vaccine_research/diseases/diarrhoeal/en/index4.html (accessed 17 September 2012)
112. Quadri F, Svennerholm A-M, Faruque ASG et al (2005) Enterotoxigenic *Escherichia coli* in Developing Countries: Epidemiology, Microbiology, Clinical Features, Treatment, and Prevention. Clin Microbiol Rev 18:465–483

113. Tsuji T, Honda T, Miwatani T et al (1985) Analysis of receptor-binding site in Escherichia coli enterotoxin. J Biol Chem 260:8552–8558
114. Tiwari S, Verma P, Singh P, Tuli R (2009) Plants as bioreactors for the production of vaccine antigens. Biotechnol Adv 27:449–467
115. Benner G, Andrews G, Byrne W et al (1999) Immune response to Yersinia outer proteins and other *Yersinia pestis* antigens after experimental plaque infection in mice. Infect Immun 67:1922–1928
116. Du Y, Rosquist R, Forsberg A (2002) Role of fraction 1antigen of *Yersinia pestis* in inhibition of phagocytosis. Infect Immun 70:1453–1460
117. Welkos S, Pitt ML, Martinez M (1998) Determination of the virulence of the pigmentation-deficient and pigmentation-/plasminogen activator-deficient strains of *Yersinia pestis* in non-human primate and mouse models of pneumonic plague. Vaccine 20:2206–2214
118. Heath D, Anderson G, Mauro J et al (1998) Protection against experimental bubonic and pneumonic plaque by a recombinant capsular F1-V antigen fusion protein vaccine. Vaccine 16:1131–1137
119. Williamson ED, Vesey PM, Gillhespy KJ (1999) An IgG1 titer to the F1 and V antigens correlates with protection against plague in the mouse model. Clin Exp Immunol 116:107–114
120. Alvarez M, Pineyerd H, Crisantes J et al (2006) Plant-made subunit vaccine against pneumonic and bubonic plaques orally immunogenic in mice. Vaccine 24:2477–2490
121. Ni M, Cui D, Einstein J (1995) Strength and tissue specificity of chimeric promoters derived from the octopine and mannopine synthase genes. Plant J 7:661–676
122. HPV Vaccine: State Legislation and Statutes (Updated 2012) www.ncsl.org/issues…/hpv-vaccine-state-legislation-and-statutes.aspx
123. Munoz N, Bosch FX, de Sanjose S (2003) Epidemiologic classification of human papillomavirus types associated with cervical cancer. N Engl J Med 348:518–527
124. Zheng Z-M, Baker CC (2006) Papillomavirus genome structure, expression, and post-translational regulation. Front Biosci 11:2286–2302
125. Yim E-K, Park J-S (2005) The Role of HPV E6 and E7 Oncoproteins in HPV-associated Cervical Carcinogenesis. Cancer Res Treat 37:319–324
126. Massa S, Franconi R, Brandi R et al (2007) Anti-cancer activity of plant-produced HPV16 E7 vaccine. Vaccine 25:3018–3021
127. Ergonul O (2006) Crimean-Congo haemorrhagic fever. Lancet Infect Dis 6:203–214
128. ECDC Meeting Report (2008) Consultation on Crimean-Congo haemorrhagic fever prevention and control. Stockholm 2008. www.ecdc.europa.eu
129. Ghiasi SM, Salmanian AH, Chinikar S et al (2011) Mice orally immunized by a transgenic plant expressing the glycoprotein of Crimean-Congo hemorrhagic fever virus. Clin Vaccine Immunol 18:2031–2037
130. Whitehouse CA (2004) Crimean-Congo hemorrahagic fever. Antiviral Res 64:145–160
131. Spik K, Shurtleff A, McEnlroy AK et al (2006) Immunogenicity of combination DNA vaccines for Rift Valley fever virus, tick-borne encephalitis virus, Hantaan virus, and Cimean-Congo hemorrhagic fever virus. Vaccine 24:4657–4666
132. Turnbull P (1992) Anthrax vaccines: past, present and future. Vaccine 9:533–539
133. Aziz MA, Singh S, Kumar PA et al (2002) Expression of protective antigen in transgenic plants: a step towards edible vaccine against anthrax. Biochem Biophys Res 299:345–351
134. Flick-Smith H, Eyles J, Hebdon R (2002) Mucosal or parenteral administration of microsphere-associated *Bacillus anthracis* protective antigen protects against anthrax infection in mice. Infect Immun 70:2022–2028
135. Brossier F, Weber-Levy M, Mock M et al (2000) Role of Toxin Functional Domains in Anthrax Pathogenesis. Infect Immun 68:1781–1786
136. Chichester J, Musiychuk K, de la Rosa P et al (2007) Immunogenicity of subunit vaccine against *Bacillus anthracis*. Vaccine 25:3111–3114

Adv Biochem Eng Biotechnol (2013) 134: 135–154
DOI: 10.1007/10_2013_180

Published Online: 6 March 2013

Metal Uptake and Nanoparticle Synthesis in Hairy Root Cultures

Zahwa Al-Shalabi and Pauline M. Doran

Abstract Hairy roots are a convenient experimental tool for investigating the interactions between plant cells and metal ions. Hairy roots of species capable of hyperaccumulating Cd and Ni have been applied to investigate heavy metal tolerance in plants; hairy roots of nonhyperaccumulator species have also been employed in metal uptake studies. Furnace treatment of hairy root biomass containing high concentrations of Ni has been used to generate Ni-rich bio-ore suitable for metal recovery in phytomining applications. Hairy roots also have potential for biological synthesis of quantum dot nanocrystals. As plant cells intrinsically provide the confined spaces needed to limit the size of nanocrystals, hairy roots cultured in bioreactors under controlled conditions are a promising vehicle for the manufacture of peptide-capped semiconductor quantum dots.

Keywords Hairy roots · Heavy metal · Hyperaccumulator · Nanoparticle · Phytomining · Phytoremediation · Quantum dot

Contents

Z. Al-Shalabi
School of Biotechnology and Biomolecular Sciences, University of New South Wales, Sydney NSW, 2052, Australia

P. M. Doran (✉)
Faculty of Life and Social Sciences, Swinburne University of Technology, PO Box 218Hawthorn, Melbourne VIC, 3122, Australia
e-mail: pdoran@swin.edu.au

1 Introduction

Plants have been studied extensively for their ability to absorb and accumulate metals from the environment. Minerals are an essential component of plant nutrition; several heavy metals such as Cu, Fe, Mn, Ni, and Zn play important roles in enzyme reactions, membrane function, and gene expression [59]. Accordingly, mechanisms of metal homeostasis and regulation of metal uptake and accumulation are of key importance in our understanding of plant biology, physiology, and metabolism. Heavy metals at high concentrations can have severe detrimental effects on cellular function and often pose an environmental and ecological threat. Depending on the species and culture conditions, plants exhibit a broad range of responses to elevated metal concentrations, from toxic reaction to varying levels of tolerance.

Several applications of commercial interest focus on the interaction of plants with metal ions and processes of biological mineralization. Opportunities exist for the development of new technologies in areas such as

- Phytoremediation
- Phytomining
- Biofortification
- Nanoparticle synthesis.

Metals such as Cd, Co, Cu, Mn, Ni, Pb, and Zn are released into the environment from a range of sources including fertilizers, sewage sludge, industrial effluent, mining, military operations, and fossil fuel combustion. Metal contamination of soils and waterways represents a health risk to humans, animals, and ecosystems and can reduce agricultural production by affecting soil fertility and crop yield. Effective, low-cost strategies are required for removing or reducing the bioavailability of metals in the environment and to remediate polluted land, groundwater, sediments, rivers, and lakes.

Of the various approaches that have been proposed for phytoremediation [14, 46], phytoextraction has received the most attention and has the greatest potential for effective pollutant removal. The general principle of phytoextraction is that plants use their roots to take up metals from the soil and then accumulate the metals in their leaves. Bioconcentration of the contaminant is thus achieved; harvesting of the aerial parts of the plants allows effective removal of the metals from the environment. An extension of phytoremediation is phytomining, which involves the economic recovery of accumulated metals from plant biomass using processes such as ashing, smelting, or liquid extraction.

Biofortification of food crops with mineral trace elements addresses a different problem: the need to overcome nutritional deficiencies in humans and animals caused by inadequate intake of micronutrients [63]. Many factors that affect the feasibility of phytoremediation and phytomining, such as the bioavailability of metals in the rhizosphere and translocation of metal ions from the roots to the shoots, are also important in biofortification schemes. Additional issues such as the effective distribution of minerals to the edible tissues of crops such as seeds are also relevant.

The ability of plants to take up and detoxify metals can be further exploited for the synthesis of nanocrystals in living plant tissues [25]. Au and Ag nanoparticles of dimensions 1–200 nm have been produced using a wide range of plant species [33]; these particles have applications in medicine, catalysis, chemical analyses, DNA detection, biosensors, electrodes, and electrical coatings. Quantum dot nanocrystals with semiconductor properties are also potential candidates for phytosynthesis after uptake of metal ions by plant cells and intracellular linkage and stabilization with phytochelatin peptides. The production of valuable metal-based nanocrystals in plants is a relatively new area of investigation involving input from a diverse range of scientific disciplines including chemistry, biology, medicine, engineering, and materials science.

The aim of this work is to examine the role of hairy root cultures in studies of plant cell responses to heavy metals. Hairy roots have been employed to investigate Cd, Cu, Ni, Pb, and U uptake and accumulation, and to demonstrate Ni recovery from metal-laden plant biomass. Synthesis of nanocrystals with quantum dot properties is a relatively new application for hairy root cultures; here we present initial results for CdS nanoparticle production using tomato hairy roots.

2 Benefits of Plant Tissue and Hairy Root Cultures in Metal Uptake Studies

Experimental investigation of metal uptake and tolerance in plants requires a convenient and reliable study system. Although most work in this area has been carried out using whole plants grown either in soil or hydroponically, there are advantages associated with alternative plant-based cultivation systems. This is especially true when the aim of the study is to examine the mechanisms and metabolic responses of plants rather than their agronomic characteristics. Accordingly, there is a role for plant cell and tissue culture in the development of phytotechnologies and related research [18].

The time required to carry out experimental investigations is often substantially reduced using tissue cultures rather than whole plants, which have a limited lifespan and must be replaced and re-established after each experiment. In vitro culture allows indefinite propagation of cells and tissues, thereby avoiding the effects of variability between individual plants. Experimental conditions in tissue culture are more easily controlled than for soil-growing plants, particularly with regard to nutritional parameters, phytohormone levels, and other aspects of the biochemical environment, as extracellular complexation of nutrients and other substrates is minimized. This, together with the relative homogeneity of cultured cells and organs compared with whole plants, facilitates the standardization of experimental procedures and improves the reproducibility of results. Axenic conditions in tissue culture also allow metal uptake to be studied without interference from microbial activity in the soil; the responses and metabolic capabilities of plant cells can thus be distinguished from those of microorganisms normally present in the rhizosphere or plant tissues [34, 60].

As an experimental tool, hairy roots offer additional features and benefits compared with de-differentiated plant tissues such as callus and suspended cells. Greater genotypic and phenotypic stability [1, 22] is a significant advantage: over time, hairy roots often prove to be a more reliable and reproducible system than de-differentiated plant cells. Hairy roots also have simpler culture requirements than plant callus or suspension cultures as exogenous growth regulators are not required. Experiments with hairy roots allow the metal accumulation capacity and tolerance properties of roots to be characterized without translocation effects; this can help determine, for example, whether translocation is necessary for metal detoxification, and the location of essential enzymes. Because hairy roots are themselves the products of genetic transformation of plant cells with bacterial DNA, further genetic modification via the Ri (root-inducing) plasmid of *Agrobacterium rhizogenes* is relatively straightforward for the introduction of improved metal accumulation traits. Hairy root cultures are also a useful tool for screening genetic transformants prior to regeneration of whole plants with enhanced properties.

Although hairy root cultures offer many advantages for investigation of metal uptake and tolerance in plants, it is important to recognize the limitations associated with the use of in vitro study systems. The conditions applied for in vitro culture of plant tissues are very different from those experienced by field-grown plants; this can introduce into experimental investigations additional variables such as medium composition, oxygen supply, and tissue morphology that affect the response of the cells. Accordingly, extrapolation of results for metal uptake and tolerance obtained using tissue cultures to mature plants in the field requires careful consideration. For environmental applications, hairy roots can be used to obtain useful information to guide and inform subsequent whole-plant trials.

Because hairy root cultures are maintained under sterile conditions and are easily destroyed by microbial contamination, they cannot be used directly in large-scale phytoremediation or phytomining operations to treat soil or liquid waste. They may, however, be applied as in vitro production systems for synthesis of valuable metal-based nanoparticles. Because the application of toxic heavy metals such as Cd to field-grown plants would be prohibited due to the serious environmental consequences of Cd release, in vitro cultures carried out under controlled conditions are an attractive alternative for biological production of nanocrystals.

3 Phytoremediation Studies with Hairy Roots

Over the last decade, there has been considerable interest in the exploitation of plants as agents for phytoremediation of metal-contaminated land and waterways. Theoretically, phytoremediation represents a low-cost, low-impact approach to environmental clean-up, as the use of plants growing in situ does not require the removal of contaminated soil and the consequent destruction of its physical and biological properties. In practice, however, several factors have restricted the application of phytoremediation methods [14, 57]; these include the limited soil depth that can be penetrated by plant roots and the long periods of time required to achieve adequate reductions in metal concentration. Some of these barriers can be alleviated by increasing the efficiency of metal uptake and accumulation in plant biomass; the commercial feasibility of phytoremediation may also be improved if it can be linked to other phytotechnologies such as biofortification, bioenergy and biochar production, and phytomining [14]. Of key importance is the availability of plant species with the ability to accumulate and tolerate high concentrations of metal ions.

3.1 Metal Hyperaccumulation

Across the plant kingdom, different species exhibit a wide range of metal storage and tolerance capabilities. Of particular interest for phytoremediation are approximately 450 plant species that have been identified as metal "hyperaccumulators" [58]. These rare plants are capable of taking up and storing elevated concentrations of Cd, Cu, Mn, Ni, Zn, and other metal elements at levels 100- to 1,000-fold higher than those found in nonhyperaccumulators [49], without suffering toxicity symptoms or cell damage.

Whether or not a particular plant species is classified as a hyperaccumulator depends on the concentration of metal stored in its aerial parts. When grown in native soil without symptoms of metal poisoning, accumulation in the dry leaves of >10 mg g^{-1} (1 %) Mn or Zn, >1 mg g^{-1} (0.1 %) As, Co, Cr, Cu, Ni, Pb, Sb, Se, or Tl, or >0.1 mg g^{-1} (0.01 %) Cd qualifies a species as a hyperaccumulator of that metal [49, 58]. Most plants classified in this way are capable of hyperaccumulating only one metal. This implies that hyperaccumulation mechanisms are largely metal-specific; however, four plant species are recognized as hyperaccumulators of both Cd and Zn [58]. More than 75 % of hyperaccumulator species hyperaccumulate Ni. No hyperaccumulators of radionuclides have been identified so far. About 25 % of known hyperaccumulators belong to the Brassicaceae family; the genera *Alyssum* and *Thlaspi* are particularly rich in hyperaccumulating species [49, 58].

The mechanisms of hyperaccumulation for different metal elements are not completely understood. A range of processes involving enhanced metal uptake, root–shoot translocation, vacuolar storage, detoxification, and tolerance is involved [58, 63]. In some plants hyperaccumulation is a constitutive trait, whereas in others it occurs only in populations adapted for growth in the presence of high metal concentrations. Microarray and transcriptomic analysis has been used to identify genes that are expressed at significantly greater levels in hyperaccumulators than in nonhyperaccumulators (reviewed in [58]); however, the function of many genes found to be upregulated in hyperaccumulator species remains unknown. Further work to combine the enhanced metal uptake and tolerance properties of hyperaccumulator plants with high biomass production is needed to improve the technical and commercial feasibility of phytoremediation.

3.2 Hairy Root Studies of Nonhyperaccumulator Species

The first studies of metal uptake in plants using hairy roots as a model system were conducted with the nonhyperaccumulator species, *Nicotiana tabacum*, *Beta vulgaris,* and *Calystegia sepium* [39]. The aim of the investigation was to assess the bioavailability of Cd in sewage sludge. Levels of Cd accumulation in the roots varied with plant species and sludge source; growth of *B. vulgaris* hairy roots was found to be highly sensitive to Cd. The lower concentrations of Cd accumulated by tobacco hairy roots compared with whole tobacco plantlets were attributed to the absence of transpiration in the hairy root cultures. Further investigation of Cd accumulation was carried out using *Solanum nigrum* hairy roots [36, 37]. From medium solutions containing 2–50 mg L^{-1} Cd provided as $Cd(NO_3)_2$, Cd was taken up rapidly during the first 0.25 h of exposure and then more slowly over a further 20–50 h [37]. Growth of the roots was almost completely inhibited at 50 mg L^{-1} Cd; at this concentration, the roots became dark in color and began to callus.

Cd uptake and tolerance were studied using hairy roots of *Rubia tinctorum* [38]. Roots exposed to 100 μM Cd in the form of $CdCl_2$ were saturated with Cd after 1 day of exposure; however, further Cd accumulation occurred as the roots grew. When roots harvested during the 14-day culture period were homogenized, the proportion of accumulated Cd recovered from the supernatant fraction increased from 33 to 77 % with increasing culture time, indicating a change in subcellular Cd distribution.

The response to Cd of the endangered plant species, *Adenophora lobophylla*, was investigated using hairy root cultures [62]. The properties of *A. lobophylla* were compared with those of the nonendangered but related plant, *A. potaninii*, to determine if the two species exhibited different levels of Cd tolerance. Hairy root growth for both species was inhibited by 50–400 μM Cd added to the liquid medium. However, *A. lobophylla* roots were found to accumulate higher levels of Cd per mg of protein than *A. potaninii*; endogenous levels of reduced glutathione (GSH) and cysteine were also higher in *A. lobophylla* than in *A. potaninii*. Overall, the results suggested that hairy roots of the two species employed different metabolic strategies for Cd detoxification.

The effect of Cd on root growth in the presence and absence of arbuscular mycorrhizal fungi was investigated using *Daucus carota* hairy roots [27]. Cd added to the medium in the form of $Cd(NO_3)_2$ was found to inhibit root growth at concentrations of 2 and 4 mg Cd L^{-1}. However, inoculation of the roots with *Glomus intraradices* fungi reduced the inhibitory effect of Cd relative to noninoculated controls, demonstrating that mycorrhizae are effective as protective agents against Cd-induced toxicity.

Hairy roots have been used to investigate the effects of Cd, Ni, and Pb on glutathione S-transferase and peroxidase activities in *Armoracia rusticana* [44]. After treatment with 0.1 or 0.5 mM Cd, Ni, or Pb for 6 h, peroxidase activity in the roots was reduced significantly compared with untreated controls. Constitutive expression of glutathione S-transferases may also have been affected by the metal treatment and the resulting cellular stress response.

Hairy roots of *Hyptis capitata*, *Polycarpaea longiflora*, and *Euphorbia hirta* were studied for their accumulation and tolerance of Cu [41]. Whereas Cu uptake levels in short-term (9-h) experiments were similar in *H. capitata* and *P. longiflora* hairy roots, the Cu content of *E. hirta* roots was lower than for the other species. When *H. capitata* hairy roots were cultured in longer-term (28-day) experiments with 20 ppm Cu and an equimolar concentration of disodium ethylene diaminetetraacetate dihydrate (EDTA), growth of the roots was not significantly affected relative to untreated controls. Cu uptake by *H. capitata* roots was biphasic with initial rapid accumulation followed by a slower increase in Cu content.

Rhizofiltration of U has been studied using hairy roots of *Brassica juncea* and *Chenopodium amaranticolor* [20]. Root growth was reduced when U in the form of $UO_2(NO_3)_2$ was added to the culture medium at concentrations of 1,000–5,000 μM. Growth of *B. juncea* was affected to a much greater extent than *C. amaranticolor*; U uptake was two- to four-fold higher using *B. juncea* hairy roots compared with *C. amaranticolor*.

3.3 Hairy Root Studies of Hyperaccumulator Species

Thlaspi caerulescens is a small, low-growing plant native to Europe and the United States. It is a hyperaccumulator of Cd and Zn. *Alyssum bertolonii* is native to the Mediterranean region and hyperaccumulates Ni. Both species are members of the Brassicaceae family and have been studied extensively in phytoremediation research. Hairy root cultures have been used to investigate Cd tolerance in *T. caerulescens* and Ni tolerance in *A. bertolonii* compared with the nonhyperaccumulator, *N. tabacum* [6–8, 42, 43].

The growth and tolerance responses of *T. caerulescens*, *A. bertolonii*, and *N. tabacum* hairy roots to Cd and Ni are shown in Fig. 1. After addition of 20 ppm (178 μM) Cd in the form of $Cd(NO_3)_2$ at the time of culture inoculation, *T. caerulescens* hairy roots exhibited superior tolerance to Cd than hairy roots of *N. tabacum* (Fig. 1a). The *N. tabacum* roots turned dark brown during Cd treatment

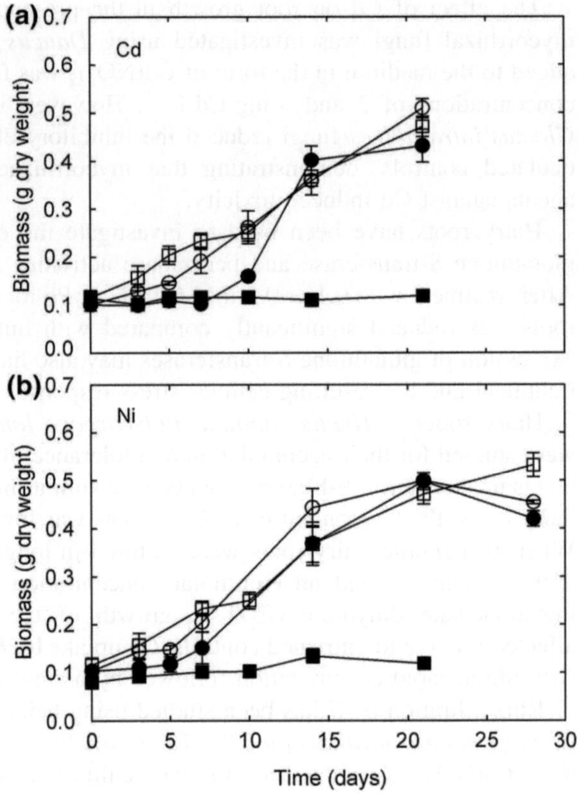

Fig. 1 Growth of hairy roots in liquid medium with and without added metal. **a** Responses to Cd: *T. caerulescens* hairy roots without Cd (*empty circle*); *T. caerulescens* hairy roots with 20 ppm Cd (*shaded circle*); *N. tabacum* hairy roots without Cd (*empty square*); and *N. tabacum* hairy roots with 20 ppm Cd (*shaded square*). **b** Responses to Ni: *A. bertolonii* hairy roots without Ni (*empty circle*); *A. bertolonii* hairy roots with 25 ppm Ni (*shaded circle*); *N. tabacum* hairy roots without Ni (*empty square*); and *N. tabacum* hairy roots with 25 ppm Ni (*shaded square*). The error bars represent standard errors from triplicate cultures (Data from [6, 7])

and growth was severely retarded. In contrast, after a lag phase of around 7 days, growth of the *T. caerulescens* roots was essentially unaffected by Cd; for *T. caerulescens*, the final biomass levels reached in the presence of Cd were similar to those in the untreated cultures. A similar overall response was found with *A. bertolonii* and *N. tabacum* hairy roots treated with Ni (Fig. 1b). When 25 ppm (426 μM) Ni in the form of $NiCl_2$ was added to the medium at culture inoculation, *N. tabacum* roots became dark brown within 7–10 days with severe inhibition of growth; in contrast, Ni treatment had a negligible effect on *A. bertolonii*. These results demonstrate that hairy roots of *T. caerulescens* and *A. bertolonii* possess superior metal tolerance properties relative to *N. tabacum*, and that the mechanisms of enhanced tolerance are functional in the absence of aerial parts of the plant such as leaves, shoots, and stems.

Active oxygen species such as superoxide anions, hydroxyl radicals, and peroxide are produced by plant cells during normal chloroplast and mitochondrial functioning and in enzyme reactions. Unless they are efficiently scavenged, these species can cause lipid peroxidation, protein degradation, enzyme inactivation, and DNA damage. Although redox-inactive metals such as Cd and Ni do not directly generate active oxygen species, exposure to metals often results in increased production of these reactive molecules, resulting in oxidative stress and damage.

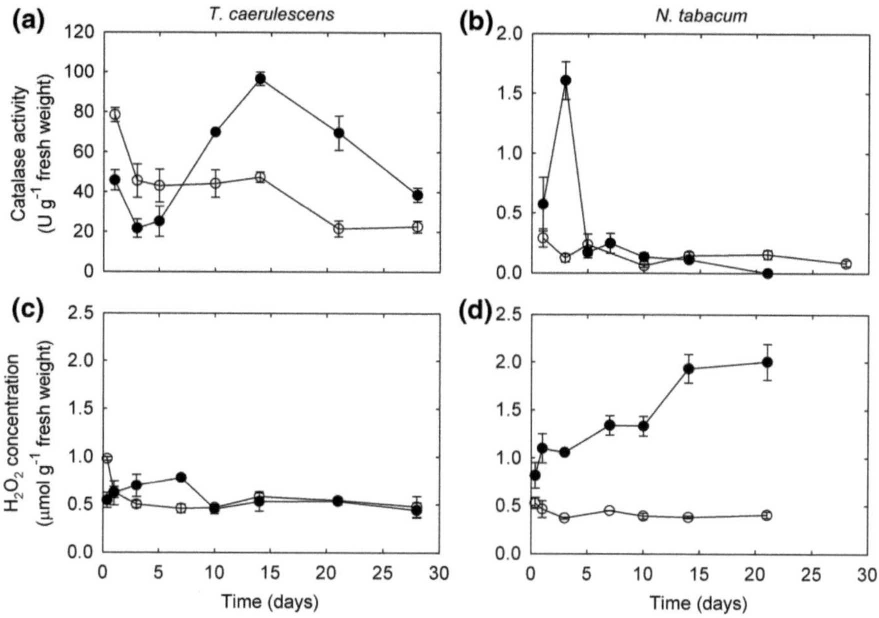

Fig. 2 Catalase activity and H_2O_2 concentration in hairy roots of the hyperaccumulator *T. caerulescens* (*left panel*) and the nonhyperaccumulator *N. tabacum* (*right panel*) grown without Cd (*empty circle*) and with 20 ppm Cd (*shaded circle*). **a**, **b** Catalase activity; and **c**, **d** H_2O_2 concentration. The error bars represent standard errors from triplicate cultures. *T. caerulescens* hairy roots contained substantially higher endogenous catalase activity than *N. tabacum*; catalase activity in *T. caerulescens* increased further with Cd treatment. H_2O_2 concentrations in *T. caerulescens* were tightly controlled and remained essentially unaffected by Cd treatment. In contrast, H_2O_2 concentrations in the *N. tabacum* hairy roots increased substantially after exposure to Cd and were significantly higher than in *T. caerulescens* (Data from [7])

For example, sustained build-up of hydrogen peroxide H_2O_2 is a typical plant cell response to toxic heavy metals, particularly Cd [54, 56]. The oxidative burst that occurs after Cd exposure has been linked to indirect inactivation by Cd of NADPH oxidases in plant cell membranes, which leads to enhanced formation of superoxide anions and H_2O_2 [23]. Excess Ni has also been found to reduce the activity of antioxidative enzymes in plant cells, thus limiting the plant's ability to scavenge active oxygen species [11]. To investigate the possibility that superior antioxidative defenses play a role in Cd and Ni tolerance in hyperaccumulator plants, the activities of antioxidative enzymes were measured in *T. caerulescens*, *A. bertolonii*, and *N. tabacum* hairy roots treated with Cd or Ni [6, 7]. The ability of these plant species to control the oxidative burst induced by metal treatment was also assessed using H_2O_2 accumulation as a marker of oxidative stress [6, 7].

Substantial differences between *T. caerulescens* and *N. tabacum* were evident in the results for catalase activity. As shown in Fig. 2a, b, maximum endogenous catalase activities without Cd were more than 270 times higher in *T. caerulescens* hairy roots than in *N. tabacum*. Although catalase activity in *N. tabacum* increased

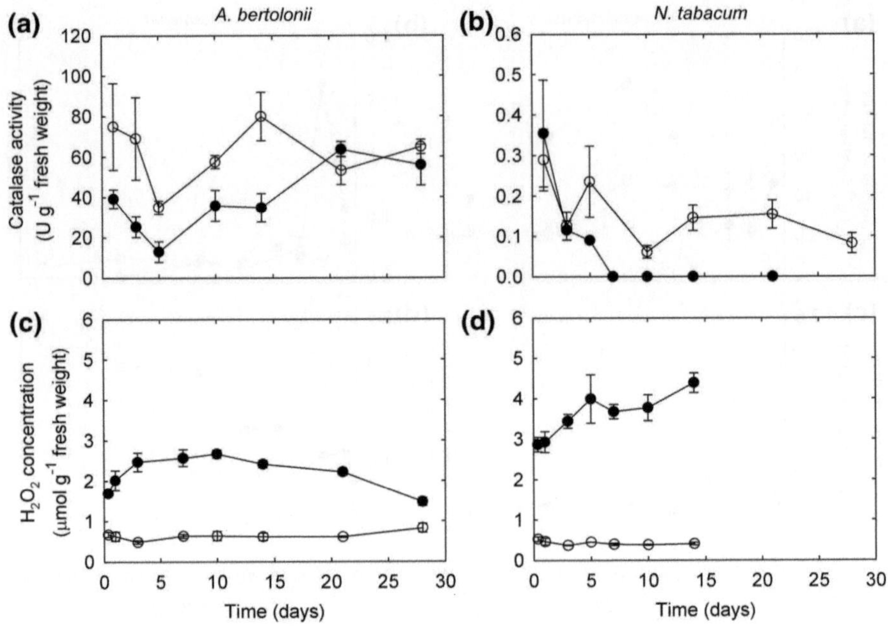

Fig. 3 Catalase activity and H_2O_2 concentration in hairy roots of the hyperaccumulator *A. bertolonii* (*left panel*) and the nonhyperaccumulator *N. tabacum* (*right panel*) grown without Ni (*empty circle*) and with 25 ppm Ni (*shaded circle*). **a**, **b** Catalase activity, and **c**, **d** H_2O_2 concentration. The error bars represent standard errors from triplicate cultures. Endogenous catalase activities were much higher in *A. bertolonii* hairy roots than in *N. tabacum*. H_2O_2 concentrations increased with Ni treatment in both species but were significantly greater in Ni-treated *N. tabacum* roots (Data from [6])

when Cd was added to the cultures, the induced levels remained substantially lower than in *T. caerulescens*. As shown in Fig. 2c, d, maximum H_2O_2 concentrations in the *T. caerulescens* roots were not significantly different with and without Cd, indicating that *T. caerulescens* exerted tight control over H_2O_2 accumulation. In contrast, Cd elicited a substantial increase in H_2O_2 levels in *N. tabacum* roots: *N. tabacum* was not as capable as *T. caerulescens* of preventing the build-up of this active oxygen species. These results suggest that *T. caerulescens* is equipped with superior antioxidative defenses compared with the nonhyperaccumulator, *N. tabacum*. The ability of *T. caerulescens* to keep H_2O_2 concentrations in check may be an important survival mechanism for this species in the presence of Cd.

As shown in Fig. 3a, b, without Ni, maximum endogenous catalase activities in *A. bertolonii* hairy roots were about 280 times greater than in *N. tabacum*. Catalase activity was not induced by Ni in either species. Endogenous H_2O_2 concentrations were similar in *A. bertolonii* and *N. tabacum* (Fig. 3c, d). After Ni treatment, even though H_2O_2 levels rose in both cultures, *A. bertolonii* hairy roots demonstrated a superior ability to control H_2O_2 accumulation compared with *N. tabacum*. These results suggest that much higher catalase activities in the hyperaccumulator roots are likely to give *A. bertolonii* an advantage over *N. tabacum* for combating Ni-induced oxidative stress.

4 Phytomining Studies with Hairy Roots

In phytomining applications, plants capable of growing in high-mineral environments are used to extract metal from the soil and concentrate it within their tissues. After the crop is harvested and dried, the biomass is treated for metal enrichment and recovery. Phytomining allows economic exploitation of low-grade surface ores or mineralized soils that are too metal-poor for conventional mining. The overall outcome is a saleable metal product and land that is more suitable for agriculture than before phytomining operations were undertaken.

Candidate sites for Ni phytomining include the large areas of ultramafic and serpentine soils at several locations around the world, including in Australia, Brazil, Canada, Italy, Russia, and Turkey. The concentration of Ni in these soils is usually within the range 0.1–0.7 % [55], which is considerably lower than the 1–2 % levels typically found in mined Ni ores [5]. When Ni-tolerant plants are used to take up Ni from the soil, leaf concentrations of around 0.2–2 % of dry biomass may be achieved [52]. These levels are enhanced further by burning or ashing the biomass to remove as much organic matter as possible and deliver a low-sulphur "bio-ore" suitable for further processing and metal recovery.

An important step distinguishing phytomining from phytoremediation is the conversion of metal-laden plant biomass into bio-ore. Relatively little work has been carried out to investigate methods for generating bio-ores or to examine whether plant-derived feedstocks are suitable for conventional mineral processing operations such as smelting, flotation, or acid leaching. However, proof-of-principle for Ni recovery and bio-ore production was demonstrated in initial studies using hairy root cultures [9]. Hairy roots of the Ni hyperaccumulator, *A. bertolonii*, were exposed to $NiCl_2$ in medium solutions to generate Ni-rich biomass. The roots were then dried, ground to a powder, and placed in a laboratory-scale horizontal-tube furnace operated at 1,200 °C under air. The resulting furnace residue contained about 80 % Ni by weight; as shown in Fig. 4, crystalline structures with a dark gray, metallic appearance were formed. Surface energy-dispersive spectroscopy (EDS) indicated that the major inorganic elements in the residue were Ni, P, Mg, and Ca (Table 1). *A. bertolonii* hairy roots thus provided a convenient model system for investigating the production of plant-derived Ni-rich bio-ore.

5 Metal Nanoparticle Synthesis Using Hairy Roots

In phytoremediation and phytomining, plants are exploited for their ability to sequester metals from the environment. Another phytotechnology that relies on the response of plant cells to metal ions is the biosynthesis of nanoparticles such as

Fig. 4 Surface morphology of Ni-rich furnace residue produced from *A. bertolonii* hairy roots treated with Ni (From [9])

Table 1 Elemental composition of furnace residue produced from Ni-treated *A. bertolonii* hairy roots

Element	%
Ni	78
P	3.2
Mg	1.1
Ca	1.0
S	0.71
Zn	0.38
K	0.31
Others	15

The elements listed are those measured at levels >0.2 % of the sample mass. "Others" refers to elements not measured, including C, H, N, and O (Data from [9])

quantum dots. This application takes advantage of metal detoxification mechanisms in plants that result in the assembly of useful metallo-organic crystal structures.

5.1 Quantum Dot Properties and Applications

Quantum dots have roughly spherical morphology and are generally only a few nanometers in size. They are composed of materials from the periodic group II–VI (e.g., CdS and CdSe), or III–V (e.g., InP). Quantum dots possess useful optical and semiconductor properties with applications in physics, biology, and medicine.

Manipulating the size and shape of quantum dots allows modification and enhancement of their optical characteristics. The size effect is related to the band-edge energy required to promote electrons from the valence to the conduction band of the nanocrystal; electron transition induces the onset of energy absorption that may be detected by UV–visible spectrophotometry. When the size of nanocrystals is reduced, the presence of fewer composing atoms results in a larger band-gap and, consequently, increased transition energy. This is manifested in blue-shifted UV absorbance peaks compared with larger nanocrystals [29]. If electrons fall from the highest to the lowest energy level within the conduction band, energy is emitted as heat in a nonradiative decay process; however, when electrons return to the valence band, energy is emitted as radiation in the form of visible luminescence or fluorescence. The energy of the luminescence peak, similar to the absorption peak, is determined by the band-edge energy and is blue-shifted as the crystal size is reduced. The ideal emission spectrum of a single quantum dot shows a narrow spectral line width where the full width at half maximum is only a few nanometers. In reality, however, because the crystal size varies within nanocrystal ensembles, the emission spectrum is broader as slightly different spectra are generated from crystals of different size [29].

The size-tuneable physical and optical properties of quantum dots are attractive characteristics for applications in light-emitting devices, computing, and photovoltaics [32]. In the biological sciences, quantum dots are used as fluorescent labels with greater brightness and higher stability compared with organic dyes. Such advantages are vital when biological reporters and probes are used in experiments conducted over extended periods of time [12, 40]. Quantum dots are also used in cell marker, cell lineage, and hybridization applications [19, 21, 26, 31]. Potential applications in the medical sciences include in tumor biology, molecular genetics, and therapeutic and diagnostic medicine [4, 28]. As the size range of quantum dots is close to that of biological macromolecules such as proteins and DNA, it is possible using these nanoparticles to penetrate biological barriers, such as the blood–brain barrier and dermal tight junctions, that other molecules of larger size cannot overcome [30].

5.2 Advantages of Biological Synthesis of Quantum Dots

During chemical synthesis of quantum dots, a high level of control must be exercised over the material components and production environment to obtain the required yield of high quality, monodisperse nanoparticles within the desired size range. An alternative to chemical manufacture is biosynthesis of quantum dots [48]. Bacteria, yeast, and plants have been found to accumulate semiconductor nanoparticles as part of their detoxification response to heavy metals. Fluorescent quantum dots can be produced using these biological agents; strategies for controlling the characteristics of biologically produced nanoparticles have also been developed [16, 17, 51, 61].

A characteristic of some biologically produced quantum dots is that they are capped during the synthesis process with phytochelatins. Phytochelatins are peptides comprising three amino acids, Glu, Cys, and Gly, arranged as γ-(Glu–Cys)$_n$-Gly, with n ranging between 2 and 11 [24]. Phytochelatins are widespread in the plant kingdom and their synthesis in plants is strongly upregulated in the presence of metal ions such as Cd [13, 24]. Because phytochelatins are rich in Cys, they can function as metal detoxification agents by binding to metal ions within cells via thiolate coordination, thus preventing the metal from causing damage to cellular components. Incorporation of acid-labile sulphur into the metal–phytochelatin complex further increases its metal binding affinity [50]. The metal binding capacity and stabilizing features of phytochelatins and related peptides have been exploited to enhance the synthesis and properties of chemically produced quantum dots [3, 47]. Although the phytochelatin-mediated pathways adopted by plants to detoxify heavy metals have been shown previously to produce nanoparticles with quantum dot properties [51], development of the technology required to utilize plant tissues as production systems for quantum dots has barely been considered.

5.3 Synthesis of Quantum Dots Using Hairy Roots

Hairy roots of *Solanum lycopersicum* (tomato) have been investigated for biological synthesis of CdS quantum dots [2]. Cd concentrations are often greater in roots than in other parts of whole plants [15, 35]; this suggests that roots may be an effective form of plant culture for nanoparticle synthesis. The choice of tomato follows reports by Reese et al. [51] of the accumulation of Cd in tomato plants in the form of phytochelatin-capped CdS crystallites.

Hairy roots were induced from tomato cv. Grosse Lisse and exposed to 100 μM Cd. To maximize the production of CdS quantum dots, different Cd dosing and culture harvesting times were examined; the optimum combination of these was defined with regard to root growth, levels of Cd accumulated in the biomass, and the ratio of inorganic sulphide to Cd in crude root extracts. UV absorbance by root extracts and their luminescence properties within the spectral ranges associated with CdS quantum dots were also examined.

Different approaches were tested for extracting CdS nanoparticles from the hairy roots, including chopping the roots in buffer and applying freeze–thaw cycles to fresh whole or chopped roots. However, Cd release using these methods was relatively low. Freeze-drying the roots and then grinding the biomass before applying freeze–thaw cycles resulted in a 10–25 % improvement in Cd recovery compared with the other extraction methods tested, as well as consistent sulphide-to-Cd ratios. Anion-exchange chromatography was applied to crude root extracts to isolate phytochelatin complexes based on the negative charge of their amino acid side chains. The recovered samples were then size-fractionated using gel chromatography.

The absorbance properties and Cd contents of 1-ml fractions collected from gel chromatography are shown in Fig. 5. The optical properties of quantum crystals depend on the particle size; however, absorption peaks at wavelengths of 280–370 nm can be expected for 2–3 nm CdS nanocrystals [10, 45, 53]. As indicated in Fig. 5, substantial absorbance peaks or shoulders were observed for hairy root gel chromatography fractions 85–115 after excitation at 280 and

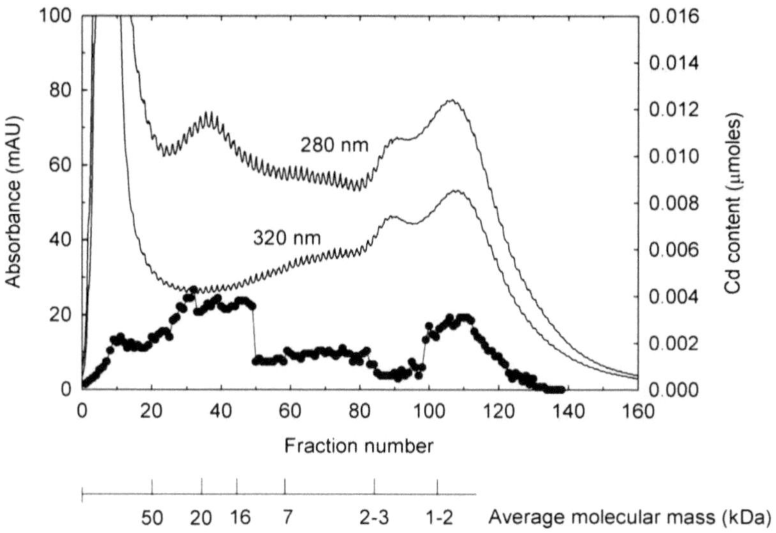

Fig. 5 Gel filtration chromatogram of tomato hairy root extract after anion-exchange chromatography for recovery of phytochelatins. The hairy roots were cultured with 100 μM Cd. Absorbance was measured at 280 and 320 nm. The Cd contents of 1-ml fractions collected from gel chromatography are also shown (*shaded circle*)

320 nm; these spectral features corresponded roughly to a peak in Cd content in fractions 100–120. The average ratio of inorganic sulphide to Cd in fractions 85–115 was 2.0. This is higher than the maximum values of 0.4 and 0.8 reported previously for Cd-treated tomato plants [51] and *Schizosaccharomyces pombe* yeast cells [50], respectively.

The luminescence emission spectra for gel chromatography fractions 86–120 are shown in Fig. 6. These spectra were obtained using an excitation wavelength of 300 nm. In gel chromatography, because larger particles are eluted before smaller particles, an increase in fraction number corresponds to a reduction in particle size. The spectra show that, as the particle size decreases, the fluorescence peaks are blue-shifted to lower wavelengths or higher-energy emissions. This property is a characteristic of CdS crystallites with quantum or semiconductor properties. The size of the particles was estimated based on their emission peak wavelength using quantum confinement theory; the results were consistent with the presence in chromatography fractions 97–116 of particles with crystal sizes of 2–20 nm.

This work demonstrates that tomato hairy roots are a promising biological means for producing CdS quantum dots. However, improvements are needed to enhance the recovery and purification of CdS nanoparticles from the biomass. Overall, the tomato hairy roots accumulated around 17 % of the Cd added to the culture medium. However, only approximately 19 % of the accumulated Cd was recovered by extraction of the root biomass and about 1.4 % of the accumulated Cd was recovered in the form of strongly fluorescing nanocrystals.

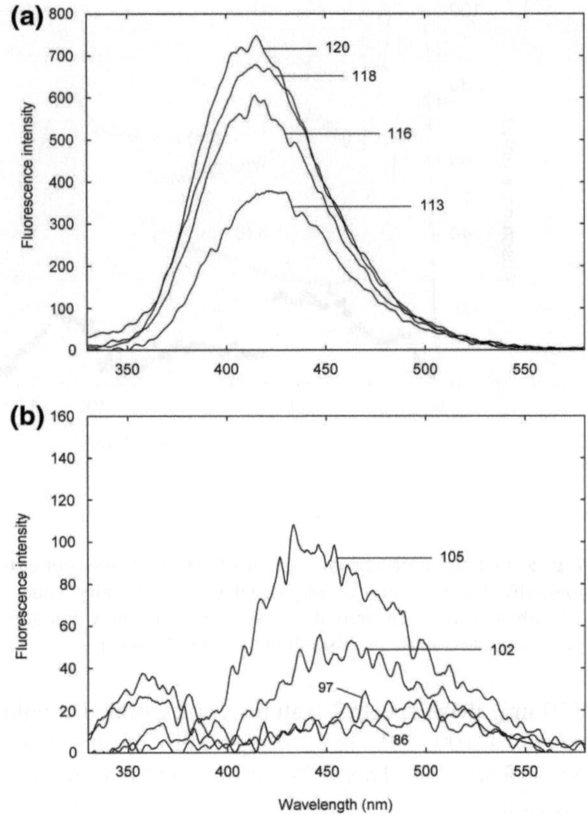

Fig. 6 Luminescence emission spectra of individual 1-ml fractions eluted during gel chromatography of tomato hairy root extract after anion-exchange chromatography for recovery of phytochelatins: **a** fractions 113–120; **b** fractions 86–105. The excitation wavelength was 300 nm. The emissions are corrected for control emissions from hairy roots not treated with Cd. The spectra were blue-shifted to higher energies as the fraction number increased, corresponding to a decrease in nanoparticle size. This relationship between particle size and emission energy is characteristic of semiconductor CdS quantum dots

6 Conclusions and Outlook

For the development of environmental technologies such as phytoremediation and phytomining, hairy root cultures are a convenient and reliable experimental system for obtaining valuable information about metal accumulation, detoxification, and tolerance mechanisms in plants. However, because the conditions experienced by whole plants in the field are very different from those applied in vitro, hairy roots represent only an initial research resource with substantial further work being required in the form of whole-plant and soil trials. Living hairy root cultures are also not a practical option for direct application in phytoremediation and phytomining operations because of their reliance on aseptic culture conditions and the absence of microbial contamination.

In contrast, hairy roots are a feasible biological system that could be applied directly for production of metal-based nanoparticles. A significant advantage of using plant cells for quantum dot synthesis is their ability to surface passivate nanocrystals with natural plant-derived organic capping layers in the form of phytochelatin peptides. Substantial further work is required to investigate the functional properties of nanoparticles formed in hairy root cultures, and to improve the efficiency of crystallite production and recovery.

Acknowledgments This work was funded by the Australian Research Council (ARC). We are grateful to Christopher Marquis and Scott Mins in the Recombinant Products Facility, University of New South Wales, for their assistance with protein purification, and to Marion Stevens-Kalceff for assistance with particle size estimation.

References

1. Aird ELH, Hamill JD, Rhodes MJC (1988) Cytogenetic analysis of hairy root cultures from a number of plant species transformed by *Agrobacterium rhizogenes*. Plant Cell Tissue Organ Cult 15:47–57
2. Al-Shalabi Z (2010) Production of cadmium sulphide quantum dots in tomato hairy root cultures. Dissertation, University of New South Wales, Australia
3. Bae W, Mehra RK (1998) Properties of glutathione- and phytochelatin-capped CdS bionanocrystallites. J Inorg Biochem 69:33–43
4. Bailey RE, Smith AM, Nie S (2004) Quantum dots in biology and medicine. Physica E 25:1–12
5. Boldt JR (1967) The winning of nickel. Methuen, London
6. Boominathan R, Doran PM (2002) Ni-induced oxidative stress in roots of the Ni hyperaccumulator, *Alyssum bertolonii*. New Phytol 156:205–215
7. Boominathan R, Doran PM (2003) Cadmium tolerance and antioxidative defenses in hairy roots of the cadmium hyperaccumulator, *Thlaspi caerulescens*. Biotechnol Bioeng 83:158–167
8. Boominathan R, Doran PM (2003) Organic acid complexation, heavy metal distribution and the effect of ATPase inhibition in hairy roots of hyperaccumulator plant species. J Biotechnol 101:131–146
9. Boominathan R, Saha-Chaudhury NM, Sahajwalla V, Doran PM (2004) Production of nickel bio-ore from hyperaccumulator plant biomass: applications in phytomining. Biotechnol Bioeng 86:243–250
10. Brus L (1986) Zero-dimensional "excitons" in semiconductor clusters. IEEE J Quantum Elect 22:1909–1914
11. Chen C, Huang D, Liu J (2009) Functions and toxicity of nickel in plants: recent advances and future prospects. Clean 37:304–313
12. Cheng F, Yu WM, Zhang XB, Ruan Y (2009) Quantum-dot-based technology for sensitive and stable detection of prostate stem cell antigen expression in human transitional cell carcinoma. Int J Biol Marker 24:271–276
13. Clemens S, Peršoh D (2009) Multi-tasking phytochelatin synthesis. Plant Sci 177:266–271
14. Conesa HM, Evangelou MWH, Robinson BH, Schulin R (2012) A critical review of current state of phytotechnologies to remediate soils: still a promising tool? Sci World J. doi: 10.1100/2012/173829
15. Conn S, Gilliham M (2010) Comparative physiology of elemental distributions in plants. Ann Bot 105:1081–1102
16. Cui R, Liu H–H, Xie H-Y, Zhang Z-L, Yang Y-R, Pang D-W, Xie Z-X, Chen B–B, Hu B, Shen P (2009) Living yeast cells as a controllable biosynthesizer for fluorescent quantum dots. Adv Funct Mater 19:2359–2364
17. Dameron CT, Reese RN, Mehra RK, Kortan AR, Carroll PJ, Steigerwald ML, Brus LE, Winge DR (1989) Biosynthesis of cadmium sulphide quantum semiconductor crystallites. Nature 338:596–597

18. Doran PM (2009) Application of plant tissue cultures in phytoremediation research: incentives and limitations. Biotechnol Bioeng 103:60–76
19. Dubertret B, Skourides P, Norris DJ, Noireaux V, Brivanlou AH, Libchaber A (2002) In vivo imaging of quantum dots encapsulated in phospholipid micelles. Science 298:1759–1762
20. Eapen S, Suseelan KN, Tivarekar S, Kotwal SA, Mitra R (2003) Potential for rhizofiltration of uranium using hairy root cultures of *Brassica juncea* and *Chenopodium amaranticolor*. Environ Res 91:127–133
21. Edgar R, McKinstry M, Hwang J, Oppenheim AB, Fekete RA, Giulian G, Merril C, Nagashima K, Adhya S (2006) High-sensitivity bacterial detection using biotin-tagged phage and quantum-dot nanocomplexes. P Natl Acad Sci U S A 103:4841–4845
22. Flores HE (1987) Use of plant cells and organ culture in the production of biological chemicals. In: LeBaron HM, Mumma RO, Honeycutt RC, Duesing JH (eds) Biotechnology in agricultural chemistry, ACS Symp Ser 334. American Chemical Society, Washington, D.C., pp 66–86
23. Gallego SM, Pena LB, Barcia RA, Azpilicueta CE, Iannone MF, Rosales EP, Zawoznik MS, Groppa MD, Benavides MP (2012) Unravelling cadmium toxicity and tolerance in plants: insight into regulatory mechanisms. Environ Exp Bot 83:33–46
24. Grill E, Winnacker E-L, Zenk MH (1987) Phytochelatins, a class of heavy-metal-binding peptides from plants, are functionally analogous to metallothioneins. P Natl Acad Sci U S A 84:439–443
25. Iravani S (2011) Green synthesis of metal nanoparticles using plants. Green Chem 13:2638–2650
26. Jaiswal JK, Mattoussi H, Mauro JM, Simon SM (2003) Long-term multiple color imaging of live cells using quantum dot bioconjugates. Nat Biotechnol 21:47–51
27. Janoušková M, Vosátka M (2005) Response to cadmium of *Daucus carota* hairy roots dual cultures with *Glomus intraradices* and *Gigaspora margarita*. Mycorrhiza 15:217–224
28. Jin Z, Hildebrandt N (2012) Semiconductor quantum dots for in vitro diagnostics and cellular imaging. Trends Biotechnol 30:394–403
29. Jorge P, Martins MA, Trindade T, Santos JL, Farahi F (2007) Optical fiber sensing using quantum dots. Sensors 7:3489–3534
30. Kairemo K, Erba P, Bergström K, Pauwels EKJ (2008) Nanoparticles in cancer. Curr Radiopharm 1:30–36
31. Kloepfer JA, Mielke RE, Wong MS, Nealson KH, Stucky G, Nadeau JL (2003) Quantum dots as strain- and metabolism-specific microbiological labels. Appl Environ Microb 69:4205–4213
32. Kramer IJ, Sargent EH (2011) Colloidal quantum dot photovoltaics: a path forward. ACS Nano 5:8506–8514
33. Kumar V, Yadav SK (2009) Plant-mediated synthesis of silver and gold nanoparticles and their applications. J Chem Technol Biotechnol 84:151–157
34. Lebeau T, Braud A, Jézéquel K (2008) Performance of bioaugmentation-assisted phytoextraction applied to metal contaminated soils. Environ Pollut 153:497–522
35. Lux A, Martinka M, Vaculík M, White PJ (2011) Root responses to cadmium in the rhizosphere. J Exp Bot 62:21–37
36. Macek T, Kotrba P, Suchova M, Skacel F, Demnerova K, Ruml T (1994) Accumulation of cadmium by hairy-root cultures of *Solanum nigrum*. Biotechnol Lett 16:621–624
37. Macek T, Kotrba P, Ruml T, Skácel F, Macková M (1997) Accumulation of cadmium ions by hairy root cultures. In: Doran PM (ed) Hairy roots: culture and applications. Harwood Academic, Amsterdam, pp 133–138
38. Maitani T, Kubota H, Sato K, Takeda M, Yoshihira K (1996) Induction of phytochelatin (class III metallothionein) and incorporation of copper in transformed hairy roots of *Rubia tinctorum* exposed to cadmium. J Plant Physiol 147:743–748
39. Metzger L, Fouchault I, Glad C, Prost R, Tepfer D (1992) Estimation of cadmium availability using transformed roots. Plant Soil 143:249–257

40. Montón H, Nogués C, Rossinyol E, Castell O, Roldán M (2009) QDs versus Alexa: reality of promising tools for immunocytochemistry. J Nanobiotechnol 7:4. doi:10.1186/1477-3155-7-4
41. Nedelkoska TV, Doran PM (2000) Characteristics of heavy metal uptake by plant species with potential for phytoremediation and phytomining. Miner Eng 13:549–561
42. Nedelkoska TV, Doran PM (2000) Hyperaccumulation of cadmium by hairy roots of *Thlaspi caerulescens*. Biotechnol Bioeng 67:607–615
43. Nedelkoska TV, Doran PM (2001) Hyperaccumulation of nickel by hairy roots of *Alyssum* species: comparison with whole regenerated plants. Biotechnol Prog 17:752–759
44. Nepovím A, Podlipná R, Soudek P, Schröder P, Vaněk T (2004) Effects of heavy metals and nitroaromatic compounds on horseradish glutathione *S*-transferase and peroxidase. Chemosphere 57:1007–1015
45. Nozik AJ, Williams F, Nenadović MT, Rajh T, Mićić OI (1985) Size quantization in small semiconductor particles. J Phys Chem 89:397–399
46. Pilon-Smits E (2005) Phytoremediation. Annu Rev Plant Biol 56:15–39
47. Pinaud F, King D, Moore H-P, Weiss S (2004) Bioactivation and cell targeting of semiconductor CdSe/ZnS nanocrystals with phytochelatin-related peptides. J Am Chem Soc 126:6115–6123
48. Prasad K, Jha AK (2010) Biosynthesis of CdS nanoparticles: an improved green and rapid procedure. J Colloid Interf Sci 342:68–72
49. Rascio N, Navari-Izzo F (2011) Heavy metal hyperaccumulating plants: how and why do they do it? and what makes them so interesting? Plant Sci 180:169–181
50. Reese RN, Winge DR (1988) Sulfide stabilization of the cadmium-γ-glutamyl peptide complex of *Schizosaccharomyces pombe*. J Biol Chem 263:12832–12835
51. Reese RN, White CA, Winge DR (1992) Cadmium-sulfide crystallites in Cd-(γEC)$_n$G peptide complexes from tomato. Plant Physiol 98:225–229
52. Robinson BH, Chiarucci A, Brooks RR, Petit D, Kirkman JH, Gregg PEH, De Dominicis V (1997) The nickel hyperaccumulator plant *Alyssum bertolonii* as a potential agent for phytoremediation and phytomining of nickel. J Geochem Explor 59:75–86
53. Rossetti R, Ellison JL, Gibson JM, Brus LE (1984) Size effects in the excited electronic states of small colloidal CdS crystallites. J Chem Phys 80:4464–4469
54. Schützendübel A, Schwanz P, Teichmann T, Gross K, Langenfeld-Heyser R, Godbold DL, Polle A (2001) Cadmium-induced changes in antioxidative systems, hydrogen peroxide content, and differentiation in Scots pine roots. Plant Physiol 127:887–898
55. Sheoran V, Sheoran AS, Poonia P (2009) Phytomining: a review. Miner Eng 22:1007–1019
56. Stroinski A, Zielezinska M (1997) Cadmium effect on hydrogen peroxide, glutathione and phytochelatin levels in potato tuber. Acta Physiol Plant 19:127–135
57. Van Nevel L, Mertens J, Oorts K, Verheyen K (2007) Phytoextraction of metals from soils: how far from practice? Environ Pollut 150:34–40
58. Verbruggen N, Hermans C, Schat H (2009) Molecular mechanisms of metal hyperaccumulation in plants. New Phytol 181:759–776
59. Welch RM (1995) Micronutrient nutrition of plants. Crit Rev Plant Sci 14:49–82
60. Wenzel WW (2009) Rhizosphere processes and management in plant-assisted bioremediation (phytoremediation) of soils. Plant Soil 321:385–408
61. Williams P, Keshavarz-Moore E, Dunnill P (2002) *Schizosaccharomyces pombe* fed-batch culture in the presence of cadmium for the production of cadmium sulphide quantum semiconductor dots. Enzyme Microb Tech 30:354–362
62. Wu S, Zu Y, Wu M (2001) Cadmium response of the hairy root culture of the endangered species *Adenophora lobophylla*. Plant Sci 160:551–562
63. Zhao F-J, McGrath SP (2009) Biofortification and phytoremediation. Curr Opinion Plant Biol 12:373–380

Adv Biochem Eng Biotechnol (2013) 134: 155–160
DOI: 10.1007/978-3-642-39019-7

Index

MIX
Papier aus verantwortungsvollen Quellen
Paper from responsible sources
FSC® C105338

If you have any concerns about our products,
you can contact us on
ProductSafety@springernature.com

In case Publisher is established outside the EU,
the EU authorized representative is:
Springer Nature Customer Service Center GmbH
Europaplatz 3, 69115 Heidelberg, Germany

Printed by Libri Plureos GmbH
in Hamburg, Germany